ROUTLEDGE LIBRARY EDITIONS: SECURITY AND SOCIETY

Volume 7

MILITARY ETHICS

MILITARY ETHICS
Guidelines for Peace and War

N. FOTION AND GERARD ELFSTROM

R Routledge
Taylor & Francis Group

LONDON AND NEW YORK

First published in 1986 by Routledge & Kegan Paul plc

This edition first published in 2021
by Routledge
2 Park Square, Milton Park, Abingdon, Oxon OX14 4RN

and by Routledge
52 Vanderbilt Avenue, New York, NY 10017

Routledge is an imprint of the Taylor & Francis Group, an informa business

British Library Cataloguing in Publication Data
A catalogue record for this book is available from the British Library

ISBN: 978-0-367-56733-0 (Set)
ISBN: 978-1-00-312078-0 (Set) (ebk)
ISBN: 978-0-367-60850-7 (Volume 7) (hbk)
ISBN: 978-1-00-310072-0 (Volume 7) (ebk)

Publisher's Note
The publisher has gone to great lengths to ensure the quality of this reprint but points out that some imperfections in the original copies may be apparent.

Disclaimer
The publisher has made every effort to trace copyright holders and would welcome correspondence from those they have been unable to trace.

In memory of Nick Fotion whose grace and decency infuse every page in this work

MILITARY ETHICS

GUIDELINES FOR PEACE AND WAR

N. Fotion and G. Elfstrom

ROUTLEDGE & KEGAN PAUL
Boston, London and Henley

First published in 1986
by Routledge & Kegan Paul plc

9 Park Street, Boston, Mass. 02108, USA

14 Leicester Square, London WC2H 7PH, England and

Broadway House, Newtown Road,
Henley on Thames, Oxon RG9 1EN, England

Set in Ehrhardt
by Inforum Ltd of Portsmouth
and printed in Great Britain
by St Edmundsbury Press Ltd,
Bury St Edmunds, Suffolk.

Library of Congress Cataloging in Publication Data

Fotion, N. (Nicholas)
Military ethics.
Bibliography: p.
Includes index.
1. Military ethics.   I. Elfstrom, G. (Gerard)
II Title.
U22.F63   1986   172'42   85–18250

British Library CIP data also available

ISBN 0–7102–0182–6 (c)

Contents

Introduction

I Military forces

Most nations of the world have found it necessary to provide themselves with military forces powerful enough to inflict mass destruction on others.[1] To the impartial observer, the existence of these forces may be surprising in view of all that can be said against them. Military forces require vast expenditures of economic and human resources. Their very existence poses great burdens for the societies which nurture them. What is more, they are not merely established by nations but are all too often used by them, either to attempt to coerce others or to inflict great suffering on them. It is an unavoidable fact that war is one of the most permanent features of human life. In the twentieth century, advances in technology have made warfare increasingly destructive, while political fragmentation and social unrest have made it distressingly common.

It goes without saying that the existence and use of military forces significantly affects the lives and well-being of all the people of the world. In our view, the function of military ethics is to identify the moral issues that arise because of these forces, explain the relation of these issues to one another, and attempt to come to terms with them. The present work is devoted to these tasks. Like all works of applied moral philosophy, ours is based on a general moral theory: viz., utilitarianism. Though this theory has been much maligned lately, we believe that it offers advantages unavailable to alternative theories and that the alternatives suffer from important weaknesses. The very fact that utilitarianism is the position being constantly attacked indicates its continuing power. Our particular stance is an adaptation of the utilitarianism developed by the contemporary British philosopher R.M. Hare. His ideas have matured over a period of several decades and are most accessible in his recent work, *Moral Thinking*.[2]

Given that our basic stance is controversial, we will, without transforming this work into a tract on abstract moral philosophy, present several arguments indicating our reasons for proceeding as we do. The bulk of these arguments will be developed in later sections of this Introduction.

Before we can tend to these matters, however, we feel that it is important to discuss the nature of our topic somewhat further. One difficulty with understanding military ethics is that often it is not easy to distinguish military organizations from groups of other sorts. An armed mob, for example, is not an army, but perhaps it may become one if it remains in existence long enough and attains a stable organizational structure. Two armed and disciplined men are not an army, and neither are ten such; but a hundred may be, a thousand probably are, and ten thousand most certainly are. Police forces are distinguished from armies, though the two groups often exchange functions, uniforms, weaponry, and organizational structure. One difference is that the use of deadly force by police, as opposed to armies, is only incidental to their primary task of securing domestic order. In addition, police do not project force beyond national borders. Qualities defining military organizations thus include size, discipline, organization, and use of weapons of mass destruction capable of being projected beyond national borders. It is obvious that it will often be conceptually difficult to distinguish armies from similar sorts of organizations. For our purposes it is not necessary to make a hard and fast distinction. The moral problems concerning the use of organized destructive force will remain the same whether employed by military or near-military organizations. What is important here is not conceptual tidiness but the actions and effects of such organizations on the lives of human beings.

It is natural to believe that the only use to which military organizations can be put is for war. This is in fact false. The military is used for everything from providing domestic security to helping implement foreign policy goals to building roads and hospitals. However, what is distinctive about military forces is that they have the capacity to make war, and much of their utility for other purposes (e.g., foreign policy) derives from this capacity. But, as it happens, the conceptual boundaries of war are not clearly drawn. In fact, it is embarrassingly difficult precisely to distinguish war from armed conflicts of other sorts. A skirmish, or even a series of skirmishes, between states is not war. If, however, the skirmishing becomes earnest, and opposing sides throw in more forces to seek victory, even small engagements can evolve into war. War is in part a matter of scale, intensity, and seriousness of purpose. Wars need not even be fought between sovereign states, as in the case of guerrilla or civil war. It does seem that the opposing parties must be determined to become states even in these cases, in part because the scale and organization required to sustain a war effort can only be supported by states or state-like organizations. So wars need not be fought only between states, though only states or state-like entities have the resources to carry on activities to which we comfortably apply the term 'war'. Once more, as far as this work is concerned, it is not essential to be able to distinguish cleanly between war and near-war. The importance and the immediacy of the moral problems will remain constant, and it is these that concern us.

II *Military and other ethics*

Military ethics can be usefully compared with applied ethics in the various professions including medicine, law, business and education. A most important difference is that it guides and constrains actions that in other contexts are normally condemned in the strongest terms, viz., the intentional killing or injuring of other human beings in large numbers and the mass destruction of property. Thus, in contrast to these other fields, the most important challenge facing military ethics is to explain how its existence is justified at all. The challenge is to explain how the use of military force can even be contemplated, let alone applied in a morally justifiable manner. This is the challenge of the *pacifists*. Another challenge, from the other direction, comes from the *realists* who argue that, though wars must be fought, the brutal and violent character of such activity forestalls any hope of any moral standards successfully guiding it. These challenges are sufficiently important that we will devote the next section to supporting the view that a military ethics is possible and justifiable.

Another difference is that the other areas of ethics attend more to the relations between individuals than to institutions. Medical ethics has traditionally focused most of its attention on the relationships between doctors, patients, nurses and other health care providers. The important life and death decisions are made on this level, even though, increasingly, institutions such as hospitals, professional organizations and governments are taking part in them. For military ethics, the emphasis is reversed. In this realm it is large institutions, armies and governments, that play the important roles in structuring individual action. Nurses and physicians could function without medical institutions, but soldiers and statesmen cannot perform their characteristic activities without armies and governments. In the military, much more than in medicine, the individual is a small cog in a large machine. Military ethics is thus an ethics of institutions, and we should not presuppose that such an ethics is structured in the same way as an ethics of personal relations. It will touch the fate of huge numbers of people and often be impersonal. What is more, the important institutions of military ethics are subject to no social control or higher authority. Hospitals are regulated by other social institutions. Governments on the domestic level are, at least to some extent, responsive to the people and society they serve. None of these constraints applies to the international level, and it is in part because of this that some have argued that a genuine military ethics is impossible.

We do not deny that individual soldiers and individual citizens frequently face important moral difficulties as persons. But even these problems are distinct from other moral problems, in that they arise in extreme situations concerning matters of life and death and where individuals function under enormous pressure. It is no accident that Hare often uses military scenarios to provide examples of problems that cannot be handled by everyday morality.[3] Even physicians deal with extreme life and death cases only occasionally, and they never have to decide whether to take action to kill a normal, healthy person. Military personnel, on the other hand, are commonly faced with problems of just this sort. A further complication of the military is the great difficulty under which judgments can be made. Civilian professionals are perhaps as close a model as we are likely to find in real life of individuals suited to function as autonomous beings. They are highly educated, often possessed of vast experience in the particular situations they are likely to face, and are practiced at making judgments and acting on them. They also enjoy the respect of others. The common soldier in the field has none of these advantages. It would be naive to the utmost to expect such an individual to be equipped to display the sort of austere good judgment expected of civilian professionals.

Lastly, military ethics will differ from applied ethics of other sorts in its complexity. For one thing, there will be great variations in the kinds of moral agents found in the military realm. These will range from common foot soldiers to staff officers to governmental leaders. All of these agents will face decisions about the use of military force but, needless to say, they will do so subject to widely differing constraints and responsibilities. A related complexity is in the types of moral problems faced by these differing agents. The common soldier may worry about checking a building for the presence of civilians before firing into it, a staff officer about the morality of a strategy involving economic blockade and a government leader about stockpiling supplies of chemical warfare weaponry. One result of these differences is a further arena of complexity: viz., the differing types of moral judgments and guidelines for making them. It may be that a relatively simple moral code, adopted unreflectively, will suffice to handle most of the problems that the ordinary soldier and, perhaps, the ordinary field level officer, are likely to encounter. Their codes, furthermore, will probably be relatively rigid and specific, more like rules than like the broad principles of judgment that constitute the professional codes of ethics of physicians. But even individuals on this level are likely to face situations that are not covered by the codes. It is our belief, as we will argue later, that it is possible to train ordinary military persons to exercise some measure of critical, reflective, judgment when faced with these situations. The key is that it cannot be presumed that they will either be able or be motivated to engage in reflective moral thinking at these junctions. It is necessary that there be some form of prior training and preparation for these situations and that there be institutional structures within the military to provide support and guidance for those attempting to cope with moral quandaries. In contrast, the moral thinking of staff officers or government officials concerning questions of strategy or governmental policy cannot be rule-guided in the way the decisions of common soldiers can be. More abstract thinking and more general principles will be required.

Other types of applied ethics share the complexity of military ethics to some degree. Business ethics may be the closest. We believe, however, that military ethics is significantly more complex than the others and will reflect this in our further analysis.

III The justification of military ethics

As was noted earlier, one of the most basic problems of military ethics is to justify its very existence. In doing so we must first consider the attacks which have been made upon it.

A number of thinkers regard the destructive power of modern military apparatus to be so great as to be uncontrollable. Because of the severity of war, it appears to them that in war all moral standards are, by definition, cancelled.[4] This realist view, as it has been traditionally called, should not be confused with a view which may seem, but is not, similar to it. That is the view that moral rules have exceptions but that the exceptions will still fall within the usual standards of morality. A good example of this sort of exception is found in medical ethics. The rule that we ought to prolong life can be thought of as having an exception when the patient is in great pain and is dying gradually. However, in making an exception to the non-killing rule which permits us either to withhold treatment or even possibly do the patient in directly, we do not think of ourselves as leaving the realm of ethics. On the contrary, the exception to the rule is made by appealing to a higher rule that we ought to do all we can to relieve suffering. In this case, since the suffering is extreme and permanent, the morally best thing to do on behalf of the patient is, presumably, to terminate his life. For the realist, the opponent in a military engagement is not the object of such kindly and thoughtful consideration. He receives little or no consideration. It is as if the opponent is considered to be no more and no less than a physical object to be destroyed, used or ignored as might please his enemy. For the realist, enough of the rules of ethics have been found to have exceptions, or better yet, have been cancelled, so that war is best thought of as falling outside of the ethical realm.

There is a somewhat less radical variant of the realist position. Like the former version, it gives the enemy little if any consideration. Yet, unlike it, this view holds that in doing so we never leave ethics. The reason we do not is that, even during war, we are still operating under an ethical principle roughly expressed as 'Each society (man) for itself (himself)'.[5] Among realists there might be some disagreement as to why this principle applies between states. Some would argue that it applies at all times since ethical concepts are applicable only among those living within one society. Others would argue that this principle is applicable only when other states have forfeited their rights to consideration by committing such criminal acts as starting war in the first place.

For our purposes we do not have to choose among these variant realist theories. All of them countenance war at least in part by the process of discounting the enemy to the zero point. All, as a result, take the attitude that in war anything goes. All see it as no more wrong to crush the enemy than to crush a bug or a rock with one's foot. They differ mainly over whether looking after the interests of one's society constitutes an ethical position or not.

Pacifist theories are at the opposite extreme. Some pacifists might countenance exceptions to the killing rule. They might permit or even encourage physicians to refrain from keeping their suffering and dying patients alive. But they would not countenance exceptions that would permit killing people as a form of punishment or in self-defence. Some pacifists give utilitarian reasons for holding to these beliefs, arguing, that in the long run, less suffering takes place if we simply let others do with us as they please. Yet others argue that pacifism shows greater respect of human life than do those positions that either approve of or tolerate war. Still others ground their pacifism in religious beliefs. Whatever version of this theory, these pacifists would all take account of the interests of all humans and therefore, unlike the realists, would not ignore the humanity of the enemy.[6]

In one sense, although these two positions react to war-like attacks in opposite ways, they tend to share a common view about the hellishness of war. Both tend to view war as hellish although, even here, there are differences within agreement. Pacifists see the hellishness as a necessary aspect of war. If killing and suffering are seen as terrible in themselves, and if war accelerates these events, then war cannot but be thought of as hellish. On the other side, the realists need not be, though they probably are, committed to the notion of the hellishness of war. They probably are, since, after all, nasty things tend to happen to both sides, especially if they are evenly balanced militarily. Yet, if the enemy is weak and therefore easily crushed, war can be seen as more glorious than hellish.

In the end, in understanding these two opposing theories, it is best not to focus too much on war's hellish features. What differentiates them best is not their views on the nature of war or even the varied reasons for holding the positions which they do, but simply their responses to war. In the extreme, one says about war that anything goes while the other says that nothing goes. In the extreme, one is totally permissive about what should be done in war, the other totally impermissive.

We find neither of these families of theories satisfactory, although there need be nothing illogical or inconsistent about them. It is conceivable, for example, that those pacifists who argue that in the long run there is less suffering when we submit rather than fight are right. In some circumstances it is more than conceivable that they are right. If the other side has overwhelming military advantage over us, it may be simultaneously moral and prudent to oppose those who would argue for war. Yet it is not likely that this version of pacifism is always right. A war can be quick, decisive, *and* relieve a great amount of suffering. In fact, many argue that the war of 1971, in what is now Bangladesh, was just such a war. India's invasion, it is said, stopped the killing of hundreds of thousands of people at the hands of the Pakistani soldiers. So in its utilitarian variant, pacifism can demonstrate only that some wars are wrong. It does not, as its partisans wish, have effective utilitarian arguments to show that we ought to stay away from all wars.

Pacifist theories that rest on rights theories do not fare much better. The costs of holding steadfastly to a pacifist position in terms of rights lost can be heavy. A pacifist might say that he is willing to suffer a loss of rights in order to keep a war from starting. He might add that he does not want to 'distance' himself from other human beings by picking up a gun and using it against other human beings. However, the problem obviously goes beyond the rights he would more than likely lose. If he is a leader of a state, a powerful person in the state, or merely an individual who might have had a significant influence on the outcome of the war had he chosen to fight, his failure to defend himself is also a failure on his part to defend others. So their rights are lost by his actions and quite possibly lost without their consent or approval. Even if the pacifist could justify his own forfeiture of rights, he will have a much more difficult time justifying actions that result in the loss of rights others enjoy.

The realist theorists have their own problems. Obviously, those realists who claim that ethics has no place in war mean only that it has no place relative to the enemy. In terms of their own side, the emphasis on behaving morally is often enhanced. War brings on new duties to the citizens of the state. Not only that, at times, it raises the standards of the rules that the citizens live by on an everyday basis. If the society has a work ethic, that ethic is now practiced on a double-time basis. If loyalty and bravery are virtues, they are even more so now. Looking at the war from inside a society, it may seem that it is indulging in an orgy of ethics.

It is only with those outside, the enemy, where ethical principles are cancelled or altered to the point of being unrecognizable. Why is this so? The answer comes back quickly. 'They are the enemy in a war where the disagreements run so deep that settlement can only come about by the use of extreme violence.' The problems begin for the realist when we ask just who the enemy is. Does 'enemy' include children? Granted that it might include the enemy's munitions workers and even their transportation people. But does it include their children? Does it also include those who are permanently disabled? How about pregnant women? It seems difficult to argue that the term 'enemy' has such a broad application as to include everybody on the other side. Especially if one's moral outlook includes valuing persons because they are persons, it would be hard to develop a realist argument that would countenance indiscriminate killing.

These sorts of considerations can be extended to apply even to certain people in military uniform. If some of the enemy has surrendered and has been totally disarmed, it seems counterintuitive to say something like 'War is war' and then shoot these prisoners. It also seems morally counterintuitive to blast away at the enemy with all the heavy weapons available when it is known that he is helplessly waiting to surrender. Like the pacifist position, then, the realist position seems to represent an exaggeration. It may very well be that during certain sorts of fighting the rule 'No holds barred', which is really not a rule but a statement about the absence of rules, applies. But when the realists apply this kind of rule to all aspects of war, their position no longer sounds convincing.

These arguments against the realists, and those presented earlier against the pacifists, help us to stake out our own position. They hardly refute these two extreme positions. Any refuting or putting to rest of these positions will have to wait until our own position has been presented. That position will be between the two extremes, but more than that, it will be nowhere near either extreme. There are positions, for instance, that are not pacifist in nature but nonetheless seem to fall near it. Walzer's is a case in point.[7] Especially during war, his standards for those engaged in combat are so high that it is difficult to imagine how they can be met during the heat, fatigue, and anguish of battle. On the other side, we want to argue, many realist-like arguments that would excuse immoral behavior of those in combat are also unacceptable. One virtue of the extreme positions of realism and pacifism is consistency, since they are made up of just a few general principles with very few exceptions attached to them. In contrast, our position in the middle is quite complex. Its principles and rules will be more numerous, as will the exceptions to them. Presumably the virtue of such a middle-of-the-road position, in contrast to the extremes, is its plausibility.

IV Normative perspective

A work in applied ethics requires a general moral philosophy to provide consistency and justification for its analyses. At present, choices are divided between some form of utilitarian theory and one or another of the rights-based theories.[8] Utilitarianism has been getting a good deal of bad press lately, particularly since the publication of John Rawls's *A Theory of Justice*.[9] We believe, however, that these approaches are not superior to utilitarianism. In what follows, we will show why these rights-based theories are deficient and why the purported weaknesses of utilitarian theory are overstated.

Although rights-based theories have proliferated in the past few years, they fall into only a few main groups. One approach, adopted by Rawls himself, and by Thomas Scanlon, grounds rights on some version of contractualism.[10] Roughly, they argue that moral principles should be conceived as chosen by all relevant persons in some sort of idealized choice situation. They argue, further, that the principles chosen should include rights of one sort or another. A related group, including Charles Beitz and Henry Shue, ground their thinking about rights on what they believe to be universal human needs, both arguing that these include freedom and security.[11] Yet another group focuses on the concept of the person and develops its concepts of rights from a consideration of the requirements of the person. The most thorough development of this approach is Alan Gewirth's, though David A.J. Richards and, to a lesser extent, Ronald Dworkin and Alan Donagan work variations on this theme.[12]

In spite of their ostensible differences in approach, these theories share the characteristic of selecting some limited range of goods which, it is argued, are required by all persons. Rawls's primary goods are the set of means required to achieve one's ends, whatever they may be. Shue's basic rights, freedom and security, are those necessary for the enjoyment of whatever other rights one may have. Gewirth's freedom and well-being are those conditions necessary for the achievement of whatever goals a person may have. In these views, each person has a very strong claim to a minimum amount of these goods, and no person may have his claims overriden in order that others may have more than this minimum. In their normative content these groups of theories differ from utilitarianism in two ways. First, instead of attempting to maximize any and all goods available, they postulate a subset of goods that are of fundamental importance and mandate that our obligation is to assure some minimum supply of these for everyone. Second, in contrast to utilitarianism which, in theory, allows for the sacrifice of some individuals for the greater good of all, rights theories mandate that the interests of the individual should never be sacrificed for the benefit of all.

Problems of fundamental theoretical justification aside, these approaches fall prey to a number of practical difficulties in application. As we shall see, when these problems are met, the differences between these theories and a utilitarian theory are less imposing than might first appear. An important problem is that the various basic rights will sometimes conflict with one another. Rawls, for instance, notes that political liberty will occasionally conflict with material well-being. He then states that people in relatively wealthy countries will reasonably give greater weight to freedom while very poor peoples will probably give greater weight to material well-being.[13] He does not work out this response in detail, and it is good for the integrity of this theory that he does not. The difficulty is that, to resolve the issue, he turns to hypotheses about the reasonable preferences of individuals in different circumstances – just what the utilitarian would recommend. Rawls may respond that an important difference is that such decisions are made only under the constraints of the original position (i.e., an idealized choice situation) – which warrants that such decisions will be fair. This cannot be the solution, however, for participants in the original position are lacking precisely the information about such circumstances.[14] They do not know what sort of society they will live in or what their exact position in it will be. Rawls believes that such ignorance is necessary to ensure that decisions are fair. Yet, important moral problems such as the above arise because of special circumstances, and it seems that only a resort to reasoned preferences is capable of resolving them.

These positions are further harried because it is sometimes impossible to accommodate even the basic rights of all people. The various sorts of lifeboat and lifeboat-like examples are relevant here. The issue at hand, that of war, is particularly difficult. The decision to go to war is the decision to deprive large numbers of people of their most basic rights, many of whom have done nothing that is clearly wrong or unjust. It is difficult to see how contemporary rights theorists can deal with this issue since an appeal to rights is simply not supposed to allow this sort of thing to happen.

Rights theorists like to think that an advantage of their approach is that it accommodates the differing conceptions people have of the good.[15] Basic rights are means necessary for achieving all other goods, including plans for the good life. The task of justice is to provide people with basic rights and then allow them to pursue whatever further personal goods they may desire. What is overlooked is that the differing plans for the good life may conflict – and often do. Plans for the good life are not operative only within a private bubble of individual space but are social. Ideals compete and conflict. Further, the good life that people seek generates further interests that less critical than basic rights but which are necessary for the completion of life plans. This is an important omission since many of the conflicts that occur in life are not of the life and death sort. This holds true even in the case of war. Nations rarely initiate wars to ensure survival, though they often use the rhetoric of survival to attempt to justify their actions. Wars are often fought for ideals or for less than vital interests. Rights theories, by definition, are unable to meet these problems. Consideration of reasonable preferences, in contrast, is a natural way to begin to meet this issue.

There is nothing novel in these criticisms of rights-based theories, though there are tendencies to either ignore or paper them over in the rush to find viable alternatives to utilitarianism. Two points are worth noting however. For one thing, these problems are felt most sharply in the arena of international relations, and most especially during war. For another, there are no effective institutional means for dealing with such problems when they arise in relations between states. It is more plausible to believe that we can establish fixed principles of rights, and adjudicate among them, within societies where machinery for doing so is in place and where there is sufficient social cohesiveness to support such machinery. Thus, the deficiencies of rights-based theories are most pronounced, and the strengths of utilitarian theories most apparent, in the area of military ethics. Furthermore, as we have already suggested, practical problems of applying rights theories necessitate moving more in the direction of utilitarianism. Even so, as we will notice later, certain problems in developing utilitarian theory as a means of analyzing practical issues will require adoption of some features of the Rawlsean approach. As we have said, the particular variant of utilitarian theory we find best suited to our project is R.M. Hare's. He believes that a viable system of ethics must contain two levels of moral thinking.[16] The most common type of moral thinking is what Hare calls the intuitive. We have all been raised to think that certain types of conduct are good and others bad. We have been trained to respond with aversion to certain acts and to approve of others. We have been taught to follow a certain number of rules in an unreflective way and to feel satisfaction when we do so. This unreflective mode of proceeding is adequate most of the time, as our moral training is designed to enable us to cope in the most efficacious manner with usual situations of life. At other times, however, intuitive thinking will be inadequate. Sometimes our moral rules will conflict with one another. Sometimes we will find ourselves in circumstances so strange and novel that the rules will not apply. Also, as circumstances change, we will find it necessary to devise new rules to instill in our children.

When intuitive thinking is inadequate, we must, Hare says, move to the critical level. On the critical level we must abandon all the unreflective responses that are part of our moral training and make decisions based only on what he calls logic and the facts. 'Logic' contains two elements, 'universalizability' and 'prescriptivity'. By 'universalizability' he means that decisions we make in one situation should be made in all relevantly similar situations. By 'prescriptivity' he means that moral judgments involve commands – indications that something ought or ought not be done. The facts that Hare believes should be taken into account when making moral decisions are those having to do with the preferences of those people affected by the decisions that will have to be made.

The concrete application of Hare's moral philosophy must, we believe, have two elements. First, it will contain recommendations about what sort of rules people ought to follow and what sorts of attitudes they should be conditioned to have. Second, it will contain an identification of general considerations and usual consequences to be kept in mind when non-rule guided critical thinking must be done. A utilitarian work, as the present one, must of necessity contain general recommendations concerning possible future problems. It does not have the luxury of being able to predict exactly what factual conditions will arise or what preferences people will come to have. For this reason, it is necessary to make some assumptions about the general sorts of preferences people are likely to have and the relative weight that they are likely to place on them.

We feel confident in presuming that most people must have a strong preference to live and that they therefore have strong preferences for the means necessary for life, such as food and shelter, and, further, they have the preferences for the secure enjoyment of their lives. They do not want to live in fear of their lives or in fear of being prevented from living as they wish. We will assume that they assign a strong weight to these preferences, stronger than to other preferences they may have. As a general rule, then, actions that support the lives and security of people are morally good while those which undermine or destroy these are morally bad. Harm is to be measured in terms of damage to people's lives and security, while benefit is to be measured in terms of support for these things. These assumptions are clearly similar to those of the rights theorists. We agree with their arguments about what people generally need and strongly prefer, and we agree that these considerations are important elements of practical moral thinking. Where we disagree, on the basis of the criticisms discussed above, is with the inference that these form the basis for rights which, in Dworkin's phrase, serve as trumps over all other preferences or considerations.

There is one exception to this. People are often willing to accept great risks to their own lives and security in order to combat challenges to the well-being of those around them. Not all people freely choose to accept these risks. Still, when these challenges occur, nations often feel justified in requiring this sacrifice of some for the sake of the many. The extent to which, and the grounds on which, this can be justified will be discussed later in this study.

V Objections and responses

A cluster of the more serious objections to utilitarian schemes based on preferences are touched upon in an argument of Ronald Dworkin's. He distinguishes between personal and external preferences.[17] A person's personal preferences are for things that will affect his own well-being, such as the preference for cool drinks rather than tepid water. But people have preferences regarding the well-being of others as well. One may wish, for example, for his children to get good grades or for his rival to be passed over for promotion. These are external preferences. Dworkin's argument against preference utility is that, if external preferences are allowed to count in utilitarian calculations, the desires people have concerning the welfare of others will skew distributions. Some will receive benefits as the result of the goodwill of others while some will be harmed because of their enmity. This result may not only be inequitable, in Dworkin's view, but it may result in people suffering from such things as bigotry.

This argument, brief though it is, brings together a number of the central criticisms of preference utility. It demonstrates how this theory allows desires and values to have weight in utilitarian calculations that are thought to be morally wrong. Preferences classified as bigoted, greedy, sadistic or simply boorish are included with the rest. A number of rights theorists, Rawls in particular, believe that it is important to label certain values as morally wrong and to eliminate them.[18]

In addition, the argument about personal and external preferences illustrates how certain of the usual counter-examples to utilitarian theory are generated. If enough people hate a particular individual, it is asked, does utilitarianism justify killing him in order to satisfy those preferences? Or, if a person strongly enjoys humiliating others, does preference utilitarianism justify indulging him, provided he enjoys it enough?

Another theme, one thoroughly worked by Dworkin as well as David A.J. Richards and others, is that utilitarianism overlooks the value and the dignity of the person.[19] It accomplishes this by viewing human individuals indifferently as mere receptacles of preference satisfaction. It does not matter, so the argument goes, what the needs or goals of the individual are. Nor does it matter for the utilitarian whether persons are sacrificed in order to indulge the wishes of others or for the common good. In particular, as Richards argues, utilitarian theory does not recognize the belief, not merely that the concerns of the individual are owed consideration, but that each individual is owed certain minimal standards of treatment. This minimum, he says, should not be undermined for the sake of the common good.

Lastly, there are difficulties about why preferences should matter at all in moral reasoning. Preferences are mere brute facts about an individual, like eye color or physical shape. Some are laudabile, others are not, and many appear simply mundane. It seems natural to think that there are more important things about an individual than his preferences. Likely candidates include his freedom, his rationality, his dignity, his needs. The rights theorists argue that these candidates will provide much more suitable foundations for a moral theory than will preference. Preferences are also faulted as being difficult to tally. Given 50 people with varying preferences, how does one go about summing them or otherwise finding some means of satisfying as great a number of them as possible?

These are all-important and difficult issues. Doing them justice would require a volume in itself. In brief, however, we will argue that utilitarianism, Hare's version of it in particular, has greater resources for dealing with these problems than is commonly supposed. Our arguments develop out of an examination of the last line of criticism discussed above. This criticism emphasizes the mundane character of preferences and argues that morality should be founded on more elevated qualities. We address two questions to these critics: Why are these other qualities important or valuable? What are we to do when conflicts arise, or when not all persons can receive the same treatment? As Thomas Scanlon points out, and as Dworkin and Rawls implicitly recognize, one of the most basic problems for proponents of the new rights theories is to keep from collapsing back into one form or other of utilitarianism.[20] Rawls, for example, must go to great lengths to try to demonstrate that the parties to the original position would not choose utilitarian principles of distribution rather than the ones he proposes.[21]

The alternatives to preferences fall into two groups. The first, including freedom, reason, and needs, is important because, according to their proponents, they are necessary means for attaining whatever it is one wants. But what are these wants other than preferences? So these are valuable only insofar as they are means of preference satisfaction.

The second group includes things like dignity and respect. What is important about these is that they do not require any particular kind of treatment. To treat a person with respect is not to act towards him in one way rather than another. Thus, one can respect even a bitter enemy whom one is trying to thwart in every way and even to kill, as

Montgomery no doubt respected Rommel. Respect involves having a particular attitude towards another rather than treating him in particular ways. This is well understood by Dworkin who argues at length that treating others with equal respect is *not* incompatible with affirmative action programs that grant special treatment to certain classes of people.[22] One way of respecting people, and acknowledging their dignity, is simply to take their preferences into account in making decisions about how they are to be treated. So, concerns of this sort need not be incompatible with preference utilitarianism. It all depends on how preferences are counted. If certain people have their preferences arbitrarily ignored, they are not being treated with respect. Or, if a person suffers simply because of the enmity of others, he is not being treated with respect.

As Dworkin recognizes, Hare's requirement of universalizability in moral thinking is well suited to meet this problem.[23] Universalizability mandates that the agent in a given moral situation put himself in the place of all the other parties to the situation. He must attempt to understand the preferences each of the parties has about possible outcomes. He then considers the group of these preferences as if each of them were his own. It is as though one person had many preferences about the outcome of a particular situation and had to decide what to do based on his consideration of all of them. The agent then makes a new preference based on his reflection on the particular preferences of all the parties. So far Hare's universalizability requirement sounds like a version of ideal observer theory, and it is, but with an important qualification. It is emphasized in Hare's early writing, more than in *Moral Thinking*, that the choice made by the agent must be one he would be willing to accept were he to be in the position of any of those significantly affected by his choice.[24] This does not imply that the preferences of all parties must be satisfied equally. It does imply that the reasoning the agent uses in making a decision must be comprehensible to all rational persons. Hare would say that this is a requirement of moral reasoning itself.

Clearly Hare's version of utilitarianism shows respect for persons in the sense of taking their interests into account and in treating them as equals. Dworkin acknowledges this.[25] The problem is that Dworkin believes that utilitarianism allows disrepect by allowing the bigoted, greedy, or otherwise hateful preferences of individuals to be included in the utilitarian judgment. Thus, the claim that utilitarianism allows reprehensible preferences to count merges with the claim that it does not respect persons.

Two elements in Hare's system resist these conclusions, however. For one thing, Hare, along with Richard Brandt, argues that only prudent preferences should be considered in the utilitarian judgment.[26] Prudent preferences are those weighed against one's other preferences, both present and future, and, most importantly, must be based on beliefs that correspond to the facts. The prudent preference is designed to complement and strengthen the whole range of preferences that one has. It is quite unlikely, therefore, that bigoted preferences will survive prudent scrutiny, since the bigoted preference is one for which there is no plausible basis in fact or for which a surfeit of unnecessary malice is attached. Further, though Hare does not explicitly say so, the considered preference of the agent will be a prudent one, based on relevant fact and formed with an eye on the whole range of preferences.

The other element of Hare's system that resists disrespectful and wicked preferences is the requirement that the decision of the agent be reasonably acceptable to all the parties to the moral situation. Or, put another way, if the agent might become any of the parties to the situation, it is unlikely that he would countenance a bigoted judgment.

These considerations do not, of course, rule out bigotry in principle, and this is unlikely to satisfy those rights theorists who believe that this is just what a respectable moral theory ought to do.[27] If that is their belief, no utilitarian theory of any sort will please them. To this Hare would respond that the purpose of moral thinking on the critical level is to free ourselves of preconceptions about what is morally right or wrong. Further, he would argue that in those cases where a bigoted judgment would be justified in accordance with his system, the dispassionate observer, in full possession of the facts, would acknowledge the correctness of the decision.[28] Difficult cases require difficult decisions, ones which, under normal circumstances, we are likely to find repellent. This returns us to the problem of the counterexamples. Hare would acknowledge that sometimes utilitarian thinking will yield results that are offensive to everyday morality. In these cases, Hare would say, the difficulty is caused by the intractable nature of the cases themselves and not from flaws in utilitarian thought. *Any* moral system would encounter similar difficulties in similar situations.

Still, although we would defend Hare on this and other issues, in our view, Hare's ideas require elaboration in several ways. As was mentioned before, one of the standard criticisms of preference utility is founded on the difficulty of making calculations of utility based on preferences. Hare says simply that the agent forms a new preference based on his consideration of the collected preferences of the parties to the moral situation.[29] This enables Hare to avoid one problem. The agent does not attempt to *add* or sum the collected preferences but makes a new one. The problem is that Hare does not explain just how the agent is supposed to go about doing this. It is clear that the agent should take into account the intensity of the various preferences, so that more intense preferences should be favored over less intense ones, but this is not a great deal of help. Hare would also say that moral thinking must inescapably be a matter of judgment.[30] That is, we cannot hope to devise a mechanical procedure to compute moral problems. The agent must rely on logic and the facts, Hare says, but the combination of these two will not automatically read out a decision for us. It is for this reason that in Hare's scheme utilitarian thinking is a matter of judgment and not of calculation. All this is well and good, but it is still fair to ask for a more detailed picture of how the agent should go about transforming a disparate collection of preferences into a single moral judgment.

The moral agent is faced with a situation in which he has collected a variety of preferences from differing people. Some of this group will be mutually compatible. Others will not. Each of the preferences will be weighted in intensity by the individual who holds it. Possibly, intensity is simply fervency with which any given individual holds a particular preference. Presumably this could be measured by simply having people state the intensity of their preferences on some scale. This is problematic, however. For one thing there is the well-known factor of preference-gluttons, who strongly desire everything and would thus receive undue consideration. For another, intensities understood in this way simply do not distinguish cleanly between preferences that are greatly different in importance. People often fervently desire things that are transitory and unimportant but are cool to more weighty matters. An individual may, for example, intensely desire the latest electronic gadget but be much less concerned about his performance in school.

A simple way to meet these problems suggested in Hare's earlier writing is to define the intensity of a given preference in terms of the number of other preferences one would be willing to sacrifice in order to satisfy it. This would neutralize the preference glutton, since one cannot desire everything without sacrificing something. It would also give a clear means of distinguishing the level of importance of differing preferences for particular individuals in a way that reflects their actual weight. Furthermore, in any given situation, differing individuals will have greater or fewer interests at stake. Everyone has a preference and may even hold to these preferences with equal degrees of intensity (as defined above), but differing participants will in fact have their lives affected to different degrees by one outcome rather than another.

The moral agent will then weigh preferences in terms of two metrics, the intensity of the preference held by each party and the stake each party has in the outcome of the issue at hand. Each participant will rate preferences in terms of other preferences that they would be willing to sacrifice for the satisfaction of the one at hand. Further, each of the parties will have greater or lesser interests at stake in any given situation. Treating people as equals requires that those with the most at stake be given greater weight. The moral agent will then make a judgment about proper outcomes on which the broadest possible range of preferences are satisfied, taking account of intensity and what is at stake for each person. Let us remind ourselves once more that this must be a judgment of considering and balancing, and not of calculating with any formal rigor. For example, it would no doubt be possible to devise a means of weighing intensity of preference against a range of interests at stake. To do so, however, would be to attempt to introduce a false rigor into the deliberation, since, for Hare, moral judgments cannot simply be read off the facts. Military ethics is particularly sensitive to problems of this sort.

Since the business of the present work is with ethical codes, principles, criteria of judgments, etc., it is important to consider another objection, one specifically focused on Hare's two levels of moral thinking. Hare claims that there are only two fundamental levels of moral thinking, the intuitive and the critical, and that these levels can be kept conceptually distinct. Critics have argued that both of these beliefs are false, that there are many different sorts of moral thinking, and that it is not always easy to assign specific concepts to one level or the other. Consider rights as an example. Certain kinds of rights, particularly the legal ones, seem to fit comfortably on the intuitive level. Other sorts of rights are not so easily placed. Dworkin's right of equal concern and respect cannot be followed automatically or applied mechanically in the way that the right to one's day in court can. Indeed, some have argued that this right is disquietingly similar to Hare's principle of universalizability. So, this is a right that seems to function on the critical level rather than the intuitive. Still other rights appear to function on both levels. The right of equal treatment, one portion of Dworkin's more complex right, can function as a guide to critical thinking. But it is also used by ordinary people as a rule of intuitive response in political sloganeering and in impulsive judgments of what is fair and just. Particular kinds of moral rules, such as 'Never tell lies', that are simple and meant to be followed mechanically, appear to be suitable for use only on the intuitive level. Others, such as Dworkin's principle, can work only as guides to critical thought. Still others seem to function on both levels, while the principles of fairness are often difficult to classify.

The arena of moral thought and moral action seems more complex than Hare recognizes. This does not imply that his distinction is without value. It means only that it needs to be worked out with greater subtlety and complexity than he has done. To say that there are a variety of moral guidelines and that they may function on both levels does not rebut Hare's claim that critical and intuitive thinking are differing kinds of moral thought. What needs doing is elaborating this general insight by showing what different sorts of moral guidelines exist and how they function on the critical and/or the intuitive level. Later in this work, for example, we will discuss differing levels of policy and command decisions and several codes of military ethics. Some of the latter will serve as rules to be followed in intuitive fashion, but other codes will allow more freedom and require creative thought. Similarly, some policy decisions will follow strict guidelines and resemble intuitive conduct while others will more closely resemble pure critical thinking. To note that the distinction between critical and intuitive thinking is not a sharp and clean one, that there is a shading off from one level to the other, does not rebut the claim that the distinction is a useful one to make. Our approach, then, is to acknowledge these criticisms of Hare's two levels but argue that these insights can be accommodated within the levels in a way that will not disrupt the role that they play in Hare's thought. In fact, we believe that such modifications will enrich and strengthen his ideas.

VI Application to military ethics

This general normative stance yields a number of guidelines that aid in the development of our utilitarian military ethics. For one thing, it is clear that military action will presumably be justified only insofar as it is necessary to nurture the lives or means to life of individual persons. We will argue later that states do not have a moral standing in themselves. States have claims to our concern only insofar as they directly or indirectly affect the welfare of individual persons. A state may seek military power or prestige in international relations, but do so at the expense of the well-being of its individual citizens. In these cases, unless it can be demonstrated that these goals will eventually result in benefits to citizens which outweigh the costs, we will argue that they are unjustified.

A further implication is that the means, practice, and justification of war, as well as the use of military force, must be weighed dispassionately in terms of their contribution to human well-being. Thus the use of poison gas in warfare cannot be ruled out *a priori* as abominable or repugnant. It is important to attempt to understand what is right or wrong about using such weapons – and perhaps to acknowledge that under certain circumstances their use in war may be justified.

Hare, along with most utilitarians, argues that when making utilitarian calculations we must take into account the preferences and well-being of all parties who will be affected by our decision. Obviously, some parties will be more affected by a given decision than others. It seems plausible to assume then that those who will be most affected should have their preferences given the greatest weight, as was argued in the previous section. Complications arise when we attempt to apply these ideas to issues of military ethics. Wars, for one thing, affect large numbers of people, and some wars have permanently altered the lives of nearly all human beings. World War II is an example. It is very difficult to take account of the preferences and well-being of all humanity. Further, when nations go to war and undertake to subdue or destroy other people, they appear to be considering only the preferences of their own citizens and to count those of the enemy as zero. A different problem is that, in time of war, certain groups of people, soldiers, are singled out to bear the greatest burden of the hardships and risks of military conflict. Often they serve unwillingly and without any appreciable gain to themselves from their efforts, and without taking part in making the decision to go to war. It is plausible to think that their preferences are being ignored as well.

These issues must be handled on two levels. When rules for the conduct of war or principles that may justify war are being considered, the preferences and well-being of all human beings must be taken into account. Thus, if it is determined that certain types of weapon such as poison gas, or certain types of tactics such as carpet bombing should not be used in war, it must be because it is to the benefit of all that this be so. Or, if it is determined that wars may be fought to secure the well-being of substantial numbers of people, it must be justified by the claim that following this principle will benefit all humanity. If, using these general principles and rules, nations are able to justify going to war, then they are in fact implicitly taking account of the welfare of all. Interests of the enemy nation's citizens are not ignored but are overruled since, in this case, they are presumed to conflict with the interests of all. The citizens of Nazi Germany, many of whom were perfectly innocent and who nonetheless suffered a great deal, simply had their interests overruled by the combined welfare of humanity. It was important for everyone in the world that the Nazi policies of genocide, torture, and greedy aggression be halted.

Even the interests of Nazis, however, were taken into account by the rules of the conduct of war. If their interests had no weight at all, it would have been permissible to kill them using any means, even after they had surrendered. Or it would have been permissible to use force far in excess of what was required to achieve military goals. Development of rules of the conduct of war indicates that the interests of the citizens of enemy states, presumed to be in the wrong, do matter.

A utilitarian system cannot justify ignoring or discounting the preferences of any human being. If preferences are the facts used for making moral decisions, and if no criteria other than preferences are relevant, then preferences must be counted. So, all preferences count, but any one may be outweighed.

In a world more closely approximating the ideal than our own, all decisions made by anyone would require this global perspective. It would, that is, require a world in which there was some sort of unified governing system that was authorized to make and enforce decisions for all human beings. However, the world as it presently exists is broken down into states, and these individual states are recognized to have special responsibility for the persons residing in them. It is a system in which governments concern themselves primarily with the welfare of the people in their charge. It is a mistake, then, to make moral decisions based on the assumption that the world more closely approximates the ideal than it in fact does. We must take account of the world as it is and not as we might wish it to be. It is important to understand what this implies and what it does not.

It does imply that the governments of nation-states are justified, given present circumstances, in focusing mainly on the interests of their own citizens. They need not, in the usual case, attempt to concern themselves with the interests of all the people in the world. Individual countries do not have the recognized authority to attempt to advance the well-being of the entire human race in each and every one of their actions. Furthermore, it would be suicidal for any nation to attempt to do so. Given present circumstances, general utility is best served if nations primarily look out for themselves and reflexively fight off predators. We may well have the obligation to strive to create a better, more unified world order, but we need to acknowledge present disorder and respond accordingly.

This does not imply that nations may ignore the interests of outsiders when their policies significantly affect them. In such cases the interests of those affected must be taken into account and afforded the same concern as any citizen. Their interests may be overridden only if the interests of all those affected require it, and it may sometimes be the case that the interests of outsiders will override those of citizens. Thus, it is in the interests of the citizens of the United States that inhabitants of banana republics be exploited so that Americans may enjoy cheap produce. Insofar as US policies reinforce these conditions, Americans benefit. In cases of this sort, however, the well-being of others must weigh more heavily since they have more at stake.

Nations, then, need not make a special effort to advance the welfare of all human beings in the course of their routine activities. The interests of all need only specifically be taken into account when we are devising rules or principles of war meant to apply to all, or when we are faced with conditions that affect us. Nation-states do, however, need to take account of the interests of all persons who will be affected by particular actions or policies.

VII Methodology

Given our normative commitment, we have endeavored to devise a methodological approach that will help us to organize and understand the moral issues relevant to the use of military power. We feel that the most useful way of doing this is to structure our study across two dimensions.

The first dimension is temporal. We will consider issues that develop during peacetime, during the period immediately preceding the outbreak of war, the period of war itself, and the period following the conclusion of war. These divisions are at best rough approximations, of course. Often wars do not begin decisively with spectacular events but lurch uncertainly from relative peace to relative war. Nonetheless, they are useful because they force us to focus attention on the many important issues of military ethics that must be faced during periods of peace. In addition, decisions made at earlier stages in the cycle will have consequences in later stages. Decisions about strategy or weapons development or military training made during peacetime will alter the ways in which wars are fought and may determine the sort of peace that can be achieved afterward. Large world events stretch out over time. The character of the events will be affected by their place in the temporal cycle.

The second dimension is the social. Any given issue of military ethics is likely to affect a variety of groups of people. Just as there is a tendency in the temporal dimension to focus on war, there is a tendency in the social dimension to focus only on the military. This is shortsighted since other important groups are affected by military issues as well. It is all too easy, for example, to overlook the role of the ordinary citizen in ethical issues of peace and war. Citizens may have a moral responsibility to consider whether certain sorts of weapons should or should not be produced, or to determine whether conscription is morally inferior to voluntary enlistment.

Within the military are the distinct groups of officers and enlisted men. Apart from these is the government, as well as private industry and other social groups, such as educational, religious, scientific and medical groups. Finally, there are ordinary citizens, considered both individually and collectively. Concerns of each of these groups should be noted for enemy and third-party states, as well as for the home state. For each of the temporal segments, we will ask whether and to what extent the concerns of each of these groups are relevant to any given issue. Often, of course, issues will arise which concern only one or some of these groups. The virtue of our approach is that it will ensure that none of these groups is overlooked and that the interrelations of groups will be understood more clearly. Examining the issues in this way may also bring to light important problems that might otherwise be missed.

Given this structure of issues, our tasks will divide into several distinct types. In some instances what will be required is some set of guiding rules, the obvious example being codes of military ethics. In other cases, what will be required is a set of criteria that will aid in the making of decisions, as when we are attempting to make decisions about whether or not some types of weapons ought to be banned or whether one sort of military strategy is morally preferable to another. Finally, it will be important to develop strategies for dealing with odd or exceptional cases, such as how to decide what to do with prisoners captured far behind enemy lines. In performing each of these tasks it will be necessary to possess information. This information must consist of more than only judgments about the consequences of various acts. It must also include data on the conditions under which the acts will be performed and the state of mind of those who will perform them. 'Due concern for safety' may mean one thing on the home front, another in basic training and still something else under conditions of battle. Standards of moral conduct purporting to guide individual conduct in these differing situations, if they are to be effective and if others may reasonably hold the individual morally accountable for his conduct in these differing contexts, must take account of these different circumstances.

PART ONE:
ISSUES OF PEACETIME

CHAPTER 1
The justification of standing armies

I Costs and benefits of standing armies

The great majority of the world's nations possess standing military forces of one sort or another. One hundred and fifty-eight countries are listed in *The Military Balance 1980–1981*.[1] National spending in support of the military, as well as the size of military forces, obviously differs greatly. Fiji spent some $3.6 million on its 1,420-man army in 1978. Luxembourg managed to spend $49 million on its 660-man army. On the other end of the scale, the United States devoted $143 billion to its military force of two million persons. The largest military force is supported by the People's Republic of China. It has a force of 4.5 million people costing some $40 billion in 1978.[2] In 1980 the people of the world spent a total of some $500 billion on military forces, with the spending rate increasing between 4 and 7% per year since then.[3]

What is remarkable about this vast expenditure is that it is so often taken for granted. Even those who are critical of the arms race and the growing influence of the military generally acknowledge the need for military forces of some sort.[4] Yet, it is difficult to find a serious discussion of why military forces are required at all. This widespread acceptance of the existence of the military is remarkable in part because of all that can be said against it. The bare monetary cost of supporting military forces, as indicated above, is huge. Military expenses absorb a substantial portion of the governmental budget of many nations. In the United States, for example, military spending accounted for 21.5% of governmental spending in 1979 and 5.2% of its Gross National Product. In Britain the 1979 cost of military spending absorbed 10.3% of the government's outlay and 4.9% of its GNP. It is quite difficult to determine the amount spent on the military in the USSR though it is generally agreed that it accounts for some 11–13% of its GNP.[5] Clearly if this money were not absorbed by military spending, at least some of it would be available to satisfy human needs for food, clothing, housing, medicine, education, etc. Furthermore, the military absorbs a disproportionate amount of the world's scientific and technological expertise, not to mention large portions of increasingly scarce natural and human resources.[6] It has been frequently argued, as well, that the rate of economic growth in the production of goods and services would be greater if a lesser portion of resources were given over to the military.[7]

The existence of armaments also contributes in other, more direct, ways to human suffering. It is not unusual for repressive regimes to use military force to maintain themselves in power. In addition, at least some wars are initiated because national leaders feel confident that their military resources are superior. The actions of Hitler's Nazi Germany and, more recently, President Hussein's Iraq come readily to mind as examples. In these cases, the existence of armies and confidence in military strength was a factor in getting wars started.

While it would be an exaggeration to say that humans do not benefit from the existence of military forces, it is clear that these forces also burden them. It is doubtful that even the strongest supporters of military spending would deny that the people of the world would be better off if the military could be abolished. Armed forces, if they are to be justified at all, must find their rationale as necessary expedients in an imperfect world. The only reason usually given as to why military forces are necessary is that they provide security for individual human beings. That is, it is argued that just as police are required to protect individuals from lawless elements within a nation, so military forces are required to protect them from external disruptions of their peace and security.

However, armies serve many functions in addition to providing security for individual citizens. Many countries possessing armies are not noticeably threatened by anyone. Still other armies are so small and ineffectual that they are unlikely to seriously deter the most paltry would-be invader.[8] In these cases, rhetoric aside, the real reasons for giving support to the armed forces must lie elsewhere than a concern for security. They are not difficult to discover. Scholars have found that nations come to possess, and continue to support, their military forces for a wide variety of reasons.[9] For example, a compelling reason, particularly for new or weak nations, is simply that possessing military forces is a symbol of sovereign nationhood. In addition, armies often serve as instruments of modernization in underdeveloped nations or simply as means to keep volatile and often unemployed young men in the 18 to 25 age group occupied and off the streets for a time.[10]

II The moral issue

Often the purposes for which armed forces are used have little or no relationship to the well-being of most citizens. Samuel Huntington points out that nations with powerful military forces, those that play large and influential roles in world affairs, sometimes have an impoverished citizenry. In other, less weighty nations, such as Sweden or Switzerland, the level of well-being of the average citizen is the envy of the citizens of far greater nations, such as the USSR (and the United States, for that matter).[11] Other researchers have pointed out that there is little correlation between the average level of individual well-being in nations and the level of military power.[12]

It is one of the fundamental contentions of this work that the costs of military forces can only be morally justified if they are related in some reasonably clear-cut way to benefits for individual human beings. This is so for several reasons. For one, the costs of these forces, whether financial or otherwise, will ultimately be borne by individual citizens. More basically, benefits for the governments of nation-states are not morally significant unless related to individual welfare. Governments of nation-states are simply administrative structures composed of two elements, human individuals and the institutional relationships between them. There is no reason to think that the well-being or ill-being of these structures should be of moral concern in the way the fate of individual humans is. Governments do not have feelings, wishes, or desires in the way persons do; and they do not possess the reason and consciousness that persons have. (This does not imply that nation-states lack duties and obligations or that their acts are not subject to moral evaluation. It means only that such evaluations must apply to them in different ways than to humans.)

Of course, the individuals who are a part of government will often prosper when the government does. Its leaders, in particular, will enjoy the benefits of its power and wealth. The well-being of these persons is important, but not more important than the well-being of any other person. Benefits, or harms, that accrue to governments are of moral consequence only insofar as they can be translated into benefits or harms for particular individuals. These consequences can then be weighed against consequences of other events. In what follows, then, we will assume that the costs of military forces will only be justified if they are outweighed by benefits for individual persons.

Of all the possible justifications for maintaining a standing army, the most fundamental, and the one most readily understood as containing important benefits for particular individuals, is the argument from security. It is most basic because it refers to the very lives, freedom from attack, and access to the means of life of a people. If military forces are to be justified at all, it must ultimately be in terms of the security they provide. For this reason, we will devote the remainder of this chapter to discussing the roles a nation's military forces play in providing security for its citizens. In later chapters we will discuss the circumstances under which the military is obligated to aid the security of citizens in other nations.

III The pacifists' challenge

Intuitively, it seems that the one occasion when an individual is justified in using deadly force against another is when his own life is threatened. Self-defence is the most plausible justification that can be given for causing harm to another. Yet, the idea that self-defence justifies the use of deadly force is not uncontroversial, for pacifists deny it.[13] If the use of deadly force for self-defence is unjustified, then the creation of military forces for this purpose is unjustified as well. So it is important to consider the pacifists' argument once again and in greater detail than we did in the Introduction.

Pacifists come in various stripes. For our purposes the relevant pacifist position is the one founded on the view that the use of deadly force for any purpose is always unjustified. To keep things clear, we will refer to the individual who holds to this position in the most consistent and thorough-going fashion as the *full-bodied pacifist*. Other pacifist positions will either be less consistent than the full-bodied version, and therefore fall prey to arguments we will mention shortly; or they will be more qualified and cautious versions of the full-bodied stance, and will therefore be closer to a non-pacifist view.

We should note that the full-bodied pacifist is not committed to the view that no defensive action at all is morally justified. Various efforts to prevent harm, including passive resistance, may well be sanctioned by him. He only denies that a *violent* response directed against human life is ever appropriate.

It has been argued that pacifism falls prey to self-contradiction. The first, less sophisticated, form of this claim is that the pacifist would certainly not stand idly by while members of his family were being killed.[14] He would leap to their defence and would think it right to do so in order to prevent innocents from slaughter. But this argument would eliminate only the weak-kneed pacifist. The full-bodied pacifist would deny that it is proper to use violence either in his own defence or in the defence of anyone else, however innocent or beloved.

The second charge of contradiction seems more difficult to evade. If life-threatening acts are always wrong, it would seem that each person has the right to be free of them. But the right is empty unless it can be enforced against would-be violators. Because of this, it must be proper to use violence if that is the only way to uphold one's right against suffering violent acts. So the pacifist must claim that we have a right of freedom from violence but deny that we can enforce or defend the right, thereby making it an empty one.[15]

The full-bodied pacifist is not trapped so easily. He can simply point out that no one is justified in taking any and all measures to defend his rights. The pacifist can defend his right but can only do so by means that stop short of violence. He may well acknowledge that these efforts will be unsuccessful against a crazed or evil aggressor, but there is no reason to believe that a right is empty unless it can always be successfully defended. If that were so, there would be precious few non-empty rights.

So the full-bodied pacificist cannot readily be convicted of gross errors in logic. What can be argued, however, is that he must make important sacrifices of a moral sort in order to preserve his position. Before we tend to this argument we must examine the positive arguments that the full-bodied pacifist might use to bolster his position. He may argue, first, that the respect and dignity owed to persons entails that they should never be the objects of violent acts. He may also argue that any human life is infinitely precious and therefore cannot be sacrificed. Finally the pacifist may hold that acts of violence are infinitely wrong and cannot therefore be weighed against one another.

These arguments reduce to just one basic claim. The pacifist may claim, for example, that all acts of violence are infinitely bad. For this reason, we cannot say that any one act of life-threatening violence is more or less evil than any other, and we cannot attempt to justify any one act by claiming that it will forestall evil of greater magnitude. A plausible support for this claim is that these acts are infinitely harmful when they destroy something of infinite worth. The physical destruction of a rock, for example, cannot be wrong in itself no matter how brutally it is accomplished. What is wrong about violence is its effect on a subject. Thus, if there are beings of infinite worth, it must be infinitely wrong to destroy them. So the pacifist's third possible justification collapses into his second.

Similarly with regard to the first justification, the pacifist must explain why killing a person necessarily always involves treating him unjustly. If a person is justly killed, he is not necessarily being treated with disrespect. Kantians would claim, furthermore, that in some cases executing those who have committed crimes is the only way to show respect for them. If, however, persons have infinite value and this entails that they may never be killed, then we can understand why killing them is disrespectful. Thus, the first justification also collapses into the second.

It is not easy to find a carefully worked-out defense of the pacifist position, though any of the above may serve as a plausible basis. However, there is one further necessary element of the full-bodied pacifist position that needs to be put in place for complete understanding of this position. The pacifist must hold that there is a great deal of moral difference, indeed all the difference, between causing harm to others and passively allowing others to be harmed. If he has done all that he is morally permitted (i.e., used all means short of violence) to prevent, say, Idi Amin from murdering thousands, then those deaths are not his moral responsibility. He may, of course, feel regret and sorrow over such death, but his position cannot allow him to feel guilt.

We wish to argue, in contrast, that the full-bodied pacifist position leads to moral paralysis that should leave him with guilt. If human life is infinitely precious, then we would seem to be committed to expending all available resources and efforts to the preservation of each human life, no matter what the cost. The basis for the pacifist position is that human life is infinitely valuable. This commits him to the view that it is always wrong intentionally to kill humans. But it also forces him to hold that it must be morally wrong to make less than the supreme effort to preserve any and all human life. This has the absurd consequence that, since it is safer to stay in bed than get up in the morning, it is our duty to stay in bed. If the pacifist acknowledges that risk is a necessary part of life and that we must evaluate the risks of various courses of action, the position fails, for this is just what the pacifist initially claims we cannot do.

Paralysis results from the fact that we are all too often faced with situations in which human life will be lost, whatever options we select. Either we kill Idi Amin or thousands of others will die. In this situation the pacifist must refuse to take decisive action and acknowledge that innocent lives will be lost as a result. In cases of this sort, the pacifist must passively accept whatever result ensues – even though he may have been able to prevent suffering and death if he had taken life-threatening action.

The disturbing element of the full-bodied pacifist's position becomes apparent when we note that he would have to refuse to torture one baby to death to save thousands of other babies from being tortured to death. The act being wrong, he cannot do it. The basis of the pacifist position is found in the logic of infinite numbers. If one infinite number is added to another, the result is the same in value. If all humans have infinite value, then the life of one person is equal in value to the lives of thousands, billions, or any conceivable number of others – all being infinite in value. The loss of one person is no greater or lesser a loss than the deaths of six thousand. Thus, the choice of which event is to occur is an arbitrary one.

This is certainly a strange position, but it is not an incoherent one. The full-bodied pacifist will not be refuted by having this consequence pointed out. Neither will he deny that we properly make moral evaluations of persons and esteem some more than others. He can acknowledge that some are honest, fair and upright, while others display opposite characteristics. He may acknowledge that people should be treated differently on moral grounds. Indeed, he must hold this view if he is to have an intelligible moral position at all, for he clings to his pacifism as a moral principle. This implies that people have a duty to uphold this principle, and they are subject to censure or disapprobation if they do not. It is even permissible to take action against them to defend one's rights.

The full-bodied pacifist must hold, nonetheless, that despite all these differences in merit each human has infinite worth. It is not clear what all this implies; but, at least, it must mean that it is always wrong intentionally to kill humans. It is at this point that the pacifist lapses into incoherence. How is it that people have some infinite value which is apart from all the particular merits and demerits they possess? Why is it that we can legitimately treat people differently because of what they have or have not done in many areas but not in the area of the deprivation of life? The pacifist has no clear answers to these questions. His belief in the infinite value of persons is too vague and abstract to be of use in making moral judgments.

In contrast, the utilitarian position offers a straightforward means of dealing with matters of life and death. It is permissible to take life in order to save lives. The joys, desires, and values associated with the life of the individual represent complex, yet finite, values that can be compared with those of others. Self-defense is permissible because of the anxiety and insecurity that would result if one's life could be taken at any time, for any reasons, and also because of the deterrence it provides against other aggressive acts.

IV Individual and social defense

The demonstration that killing in self-defense is morally justified, while necessary for the justification of standing armies, is not sufficient. As was shown earlier, it is only the fate of persons within nations that is of moral significance. National military self-defense is only justified if it can be understood in terms of consequences for the lives and well-being of individual persons. From this it follows that, in principle, not all threats to the government of a nation, or even to the nation itself, understood as the common life of a people, justify an armed response.[16] Governments can be overthrown bloodlessly and cultures eroded gradually. If national military self-defense is to be justified in general terms, it must be because there are reasons for believing that in most cases an armed assault on a nation is likely to result in significant harm to the populace.

It is plausibly argued, for example, that the best way to avoid the dangers of an armed assault is to have no military force at all and to publicly announce a policy of non-violent response to an invasion.[17] The idea is that, if an invading force encounters no military opposition, it will be much less likely to make a violent entrance. This is an attractive idea in many ways. The problem with it, of course, is that recent human history does not lend it much support. There are all too many examples in this century alone of innocent and defenseless populations being treated with great brutality by armed invaders.

Going without any means of armed defense is roughly equivalent to walking unarmed through Central Park in New York City in the middle of the night. There may well be certain occasions when non-violence will be the safest response, though it seems unlikely that it would succeed as a general policy. In any case, the consequence of the failure of such a policy would be substantial.

This difficulty with the policy of non-violent response is one facet of one of the most basic practical problems of military and political affairs. It is the problem of the relation between capacities, motives and intentions. An armed invader of a defenseless nation has the capacity to wreak terrible devastation on it but may not intend to do so. Those who advocate policies of non-violent response base their position on two assumptions. One is that armed aggressors never have the motive of causing mass destruction but, instead, always have other goals such as territorial expansion, security or access to material resources. Proponents of this belief have an unlikely ally in Clausewitz who argued that aggressors would be pleased to be able to achieve their goals without fighting.[18] The second assumption is that an aggressor able to achieve his goals without violence will not develop violent intentions. The tenuous character of each of these beliefs is readily apparent, and consideration of the relation between them reveals a deeper problem. Destructive capacities are relatively stable and can be measured with ease. That is, a nation either has so many guns, tanks, etc. or it does not. If a nation does not have these capacities, this too can be known, and it can be known, as well, how long it would take to develop them and what resources are required to do so. Intentions, on the other hand, are difficult to discover with any degree of certainty and, worse, they can change instantly. A nation's motives in the sense of its interests and goals are less volatile and more readily understood than its immediate intentions, but are still shadowy. We can assume that what a given nation will do at a given time results in some fashion from the combination of its capacities, its motives and its intentions. The difficulty is that, while capacities are generally speaking both known and stable, motives are less so and intentions still less. It is, therefore, very difficult and risky to attempt to predict what any given nation will do, and the project becomes all the more hazardous once we recognize the consequences of miscalculation. A nation that has the capacity to destroy the government of another also has the capacity to destroy its citizens, and can quickly and easily develop the intention to do either. It can develop these intentions even though it has no public discernible motive for doing so. It is for this reason that, during wartime, policy analysis must focus on capacities and cannot afford to rely to any great degree on hoped-for good intentions on the part of others. Furthermore, during wartime, a belligerent nation will almost always imagine it has good motives for destroying its opponent. It is unwise, therefore, to presume that the violent intentions of invaders will spend themselves when defeating governments. As we shall see later, motives and intentions cannot be ignored; but during wartime capacities must receive the greatest emphasis.

There are yet other practical problems involved with the attempt to disentangle the governments of nation-states from the lives and well-being of their citizens. One is that outsiders will find it difficult to overthrow a government without substantially endangering the lives of large numbers of people. Surgical intervention may be possible where a government is small, weak, and thoroughly alienated from its citizens.[19] Yet, these conditions are rarely found, and when they are not, threats to governments cannot be cleanly separated from threats to citizens.

A second problem is that the security and ways of life of individual persons require a stable government as well as stable social practices to guide governmental operations. Sometimes, of course, governments need to be changed and social practices ought to be disrupted. But this cannot be a common or regular occurrence if people are to be able to live their lives in normal fashion. The experience of people in third-world countries vividly illustrates the importance of stable governmental and social institutions. Social chaos ultimately threatens life and security, even if its immediate effects do not directly do so.[20] For this reason outside aggression can be resisted not only because of its direct consequence for people's lives but also to avoid longer-term social problems. Because aggression narrowly aimed at governments will often have this broader consequence, it too may often be resisted on grounds that it threatens lives and well-being.

These considerations illustrate the general wisdom of having a rule that outside aggression will be met by an armed response. This has become a keystone of international relations. Its wisdom lies in the importance of promoting the conditions necessary for maintaining social stability and in the importance of giving pause to would-be invaders. Certainly no nation, however strong, can expect to ward off any and all external aggression; but any nation, however weak, that displays willingness to make things messy for opportunists, can expect some measure of security. This picture is complicated by the practice of uniting with other nations to provide common security. These groups face any number of complex problems, but the idea of group security is clear – and possibly even attainable given the right conditions. We will examine these problems further in Chapter 6.

States cannot kill to defend *themselves* in the way that individuals can. They can only kill to defend the lives and well-being of their citizens. Where these lives and well-being are not threatened, directly or indirectly, even defensive war is unjustified. As we have pointed out though, these situations are rarely simple and clear-cut. It is difficult to disentangle the fate of a government from that of its citizens, and it is usually unwise to rely on the good intentions of an armed aggressor. In most cases, therefore, it is plausible to presume that the lives of people will be threatened and that a counter-threat is in order.

V The problem of 'security'

Problems of self-defense, vexing though they are, do not constitute the whole or even the most complex portion of the problems we must consider. People and nations worry not only about fending off attacks in progress, but also about preventing attacks and insuring that they will be able to make an effective response to them if necessary. It is easy to see why the desire for security is nearly universal. Unfortunately, security is among the most difficult and least understood concepts of international relations. The struggle for security is responsible for many of the difficulties the world faces. The problems are of several, interrelated, sorts. There are irritating problems involved in trying to understand what security is and when one is secure. In addition, the existence of security depends on a variety of factors many of which are not easily controlled. Finally, security is necessarily a matter of degree, and so it is essential to have some idea of what degree of security can reasonably be sought.

People and nations are secure when they are not threatened by external attack and/or when they have the means to ward off any attacks that may occur. This much seems clear. Unfortunately nations may feel threats to their security from a wide variety of sources other than direct physical assault. A change of regime in a nation controlling supplies of a resource deemed vital to another nation may undermine the security of the latter even though the supplier is incapable of direct physical assault on it. Thus, one nation may be capable of threatening the security of another even though it has little or no military means of doing so. Worse, nations have a considerable degree of latitude in determining what their vital interests are and what constitutes a threat to them. The United States, for example, is notoriously volatile in its understanding of the scope of its vital interests. At certain periods in its history it has decided that its vital interests end at its borders and that nothing short of direct physical assault is a threat to it. During other periods it determines that its vital interests extend throughout the world and are vulnerable to a wide variety of threats. Both Japan and Great Britain have gone through similar cycles.

There is an ineliminable subjective element in the idea of a threat and thus of security. In many ways a nation *is* threatened to the extent that it *feels* threatened, and it is free to define the areas of interest and threat. Another element of subjectivity is that nations are not only as strong as they are, but are as strong as they appear to others. International relations is, to a large extent, a game played with mirrors and bluff.[21] Short of actual test in physical combat, a nation's genuine strength cannot easily be separated from its apparent strength. For this reason, nations go to great lengths to impress others of their prowess. Worse, nations sometimes worry more about the appearance of strength than actual military capacity. It is well known, for example, and freely admitted by all parties, that the nuclear weapons possessed by the US and the USSR are far in excess of what is required for the military task of destroying one another. The arms race between the two nations is driven in part by the desire of each to avoid being perceived as inferior relative to the other.

These subjective elements are important, then, because they imply that nations have a choice about what constitutes a threat to their vital interests and because, more importantly, they are able to choose to a large extent even what they deem their vital interests to be. Because these choices will have important consequences for the lives and well-being of individual persons, they are properly subject to moral, as well as prudential, evaluation. The game of perceptions is also a dangerous and expensive one, as is demonstrated not only by the arms race between the United States and Russia but also by all the lesser, though more volatile, arms races throughout the world. The consequences of the game are too important for either participants or observers to follow blindly. The fact that it is a game and has a subjective element undercuts the realist's belief that interests are brute facts and that participants are forced to support them at any cost or die. Still, perceptions do matter, if for no other reason than that they influence nations' actions and fears. They matter, as well, for a more important reason. In the realm of security it is impossible to separate perception cleanly from reality. It is for this reason that pacifist proponents of unilateral disarmament can rightly be accused of ignoring important reality. The game of perceptions must be controlled and understood. The problem is to determine how this can be done.

These subjective elements of security understood in the broad sense of freedom from threat to vital interests are certainly troublesome. Because they are, it is tempting to return to the core understanding of security as freedom from threat of physical assault. Unfortunately, 'being free from the threat of physical assault' is also extremely difficult to define and to apply accurately. A number of authors take the straightforwardly subjective position that people are as secure as they feel.[22] Hence, it is argued that, even though the US is spending more on defense, citizens feel even more anxious than when defense spending was lower and, therefore, they are less secure. But if this is rejected as too crudely subjectivistic, it is at least verifiable in the sense that we can measure people's attitudes.

A more austere view is that a nation is secure to the extent that it has not in fact been attacked. This too can be readily verified, and it seems to be the argument relied upon by those who claim that the US policy of nuclear deterrence has worked. The policy works, Secretaries of Defense commonly say, because the US has been free from attack since World War II. Therefore, nuclear deterrence has brought security.[23] This admirably straightforward understanding unfortunately ignores the element of threat. The United States has not felt notably secure in the last 40 years, even though it has been free of direct assault, because its citizens have felt threatened. But, once more, the notion of 'threat' has an inescapably subjective element. One is threatened if one feels threatened. So even the core definition of security has a necessary subjective element.

Another element of the general problem of security that is all too often overlooked is that military strength is only one of a number of factors that affect a nation's security. In many cases, perhaps most, military strength is not even the most important element in determining external threat. Sometimes military strength creates insecurity. The fact is that a nation with vast military resources for defending itself can as easily use them to attack others. In spite of substantial efforts to define a formula for distinguishing offensive from defensive weaponry, there has been little success in cleanly separating the two.[24] A nation that is militarily strong automatically becomes a threat to others simply in virtue of its destructive capacity, and thus invites counter-threats in turn.

The ambiguity of military prowess in providing security is well illustrated by comparing Switzerland with France. Switzerland, with only a small army, has not been attacked in hundreds of years, while France, long a military power, has been involved in numerous, often disastrous, wars in recent centuries. Clearly geography plays a central role in providing security, explaining why Switzerland is relatively secure while Poland is frightfully insecure. Ideology is also an important factor, accounting at least in part for the US and the USSR being at one another's throats. Economics is important as well, as are grievances and injustices both ancient and modern. Finally, there is the old-fashioned desire for territorial expansion well illustrated by Libya's varied adventures in northern Africa. The lesson is that a clear understanding of the security of a given nation must take into account a wide variety of factors. Military strength may be an element that adds to security, but it can also increase insecurity by making its possessor a threat to others. The role that it plays in bolstering security will depend on its interplay with these other factors.

Discussions of security are made complicated also because security is a matter of degree. There is no sharp line dividing security from insecurity. There is only a continuum of relative threat and relative vulnerability. The limit of insecurity is when one is under attack by an overwhelmingly powerful force. Even here it is difficult to know when he has reached the limit – and has plunged over the abyss to the decisive security of defeat. When one is under direct attack, he need not be insecure so long as he has an effective means of response. The limit of complete security is easier to define. It occurs at that point where one is completely invulnerable because he has nothing to lose. This point is, of course, death. Most nations and most persons are arrayed somewhere along this spectrum. The neurotic search for total security is futile and dangerous.[25] So long as nations or persons have something to lose, they are vulnerable to threats – and to that extent insecure. Furthermore, they can only approach this limit by reducing the ability of others to threaten – and thus they become more of a threat to *them* and invite a counterresponse. Clearly, relying on military strength alone will have this result, and it may be the case that other methods will have the same consequence. The search for security is not simply a question of neutralizing threats but of deciding what level of security is attainable and determining at what point the search for security becomes counterproductive.

Planning for security involves a complex array of moral and prudential factors. Moreover, since the consequences of irresponsibility, narrowness, or stupidity are so overwhelmingly great, the prudential concerns are closely linked to the moral necessity of not wasting or endangering lives needlessly. In this area as much as any other we have a moral imperative to be wise.

Usually, simple threats are less destructive of life and well-being than is direct assault. Since it is always possible that the threat will not materialize in action, that it is unintended or misconstrued, the danger is less great. This being so, less extreme measures can be justified than when warding off a direct physical assault. Those who respond to all threats as though they were actual assaults are not only acting imprudently but also acting immorally. They are endangering life and well-being in an irrational and reckless manner.

An obvious problem, however, is that threat can move all too quickly to actual assault. Sometimes, to wait until an assault begins or is clearly imminent is to wait too long, since the capacity to make an effective response is lost. Sometimes, in other words, preemptive strikes are both necessary and justified. The difficulty is knowing when they are justified. Reasonable judgments on these matters can only be made on the basis of accurate information not only about military capacities but also about the motives and intentions of those in other nations. Because the consequences for human life and well-being of making a mistake are so great, this responsibility is moral as well as prudential. Thus, an unexpected consequence of the concern for security is that there is a moral responsibility to gain a sympathetic understanding of the situation of those in other nations. But there is also another consequence here. It is that nations have the moral responsibility to allow others to gain essential information about their workings so that these other nations can make judgments about their intentions. Nations, in other words, have a duty to be much more open about their affairs than they are accustomed. Governments cannot be completely open, of course. Details of their contingency planning and tactics for war would, if widely known, lay them open to countermeasures. However, much can and should be made available to others, including governmental policies, beliefs about their security needs, their fears, and their military capacities. Often there is little harm in making many military capacities known, since it will be impossible to keep the capabilities of the weapons in service secret anyway. Another, perhaps surprising, correlate is that, in the event that this information is not openly forthcoming, other nations have a duty to uncover it – by covert means if necessary. Gathering this information on the perspectives of other nations is the analog, on the international level, of Hare's concern with putting ourselves in the place of the other in order to make moral judgments on the personal level of moral action. Prudential judgments, as well as moral ones, require this ability.[26]

Another factor contributing to the security problem is that there are several forces at work that serve continually to expand the web of security concerns and increase the area which is sensitive to threats. This point can be illustrated by two examples. It is natural and understandable for nations to desire to gain security by banding together with others in formal and informal ways. Part of the idea is simply that there is strength in numbers and another part is that nations share interests, culture or ideology in ways that make them natural allies. No doubt there is strength in numbers, but it is also the case that banding together in this way increases the area of security sensitivity for each nation. A nation in alliance with others is bound to be concerned not only by threats to its own interests but is also committed to being concerned with threats to the interests of its allies. As a result it has more to worry about. Another pressure tending to broaden the area of a nation's security sensitivity is the concern of nations to have adequate means to respond to direct assaults. Nations thereby become concerned about lines of supply and communication. Once more, this is understandable, but it greatly increases the array of issues that are perceived as security problems. A change of government in a far-off and otherwise inconsequential nation becomes a security problem if that nation happens to lie astride important shipping lanes or is an important source of supply of some strategic material.

A nation that overextends its area of moral sensitivity or wishes for complete security becomes the international equivalent of a neurotic. On this level, however, it can be more than just a nuisance, for it is likely to overreact to inconsequential matters, try to control events in areas that are not within its area of sovereignty, and stir up hostility and suspicion. No one can deny that it is difficult to find just the right balance between neurotic oversensitivity and excessive complacency. Sometimes, as before World War II, complacency was the main problem. At other times, during the height of the Cold War for instance, oversensivitity was the danger. There is no easy formula for finding this balance. In addition, the wise nation will distinguish between military threats and threats of other sorts, such as economic or ideological. It will not always be possible to separate these cleanly in practice, but it is surely the case that only an immediate military threat directly endangers life and well-being and therefore justifies an overt military response. Sometimes ideological or economic conflicts will generate military responses. When this occurs saga-cious statesmen will recognize that the ultimate solution is not military.

As a rule, then, because of the dangers of a military response, reliance on military pressure or threat will lie on the bottom of the statesman's bag of tricks. The actions that nations perform or are likely to perform depend on the interaction of their capacities, motives and intentions. Unfortunately, all too often foreign policy focuses on short-term intentions or on neutralizing or reducing military capacities. But these are not the crucial factors, nor the ones most amenable to the kinds of pressures that produce stable rela-tions. As we have noted, intentions can change almost instantly. Military capacities can only be dealt with by increasing one's own armaments, or by means of a preemptive strike, since nations rarely reduce their military voluntarily. This is not only risky but is likely to result in an arms race and increasing tensions. The key to long-range stability in relations with nations lies in focusing on their motives, their long-range interests and fears. It is only by removing the motives that nations have for conflict that their readiness to turn to military force will permanently be reduced. Motives are not per-manently altered by threats or displays of force, as several studies demonstrate. Motives can only be altered by addressing the fears, interests, and desires of other nations. These efforts, of course, will not always be successful. Sometimes conflicts of interests are ir-reconcilable. Sometimes leaders turn to military adventures because of domestic political considerations. Sometimes nations are simply greedy or irrational. Recourse to military power will always remain the option of last resort, but the wise nation will reserve it for the last resort.

Lastly, the wise nation will recognize that classifying an issue as a security concern does not automatically raise that issue to the level of highest importance, nor does that classification serve to justify any and all measures in response. Not all threats to security are equally pressing or equally matters of life and death. As was pointed out earlier, the range of sensitivity that a nation may establish to questions of security may be quite broad so that even trivial matters may come, under one rubric or another, to be classed as matters of security. Related to this is the fact that over-concern for short-term security matters can lead to long-term insecurity. A nation that is overly prickly and responds to each and every threat as though it were of the greatest importance lays the basis for its own long-term insecurity by causing others to be suspicious of it and causing them to expect an overreaction to matters of even minor concern. Also, nations do, after all, have duties, obligations, and concerns other than those of security, and sometimes these obligations will outweigh matters of security, particularly in cases where these concerns of security are relatively trivial. Most notably, nations have an obligation to concern themselves with the lives and well-being of the peoples of other nations of the world. Sometimes, in fact, as in the instance of the Nazi treatment of the Jews or the actions of the Pol Pot government in Cambodia, the moral imperative to take action may be quite strong and even substantial security concerns may have to take second place. Utilitarian morality does not justify international egoism.

We fully acknowledge, then, the great costs both material and moral of maintaining armed forces. We agree with the pacifists that the decision to go to war, even in self-defense, is a morally difficult one. But we believe that the moral costs of failing to act in self-defense are even greater. We also acknowledge the great costs of the desire for security. The effort to provide security introduces an entire new dimension of uncertainty and difficulty in world affairs. We argue, however, that if self-defense is justified, security must be as well. So while the cost of supporting military forces is great, the cost of not doing so is even greater. We are sympathetic with those who wish for a world in which this price would not have to be paid, but we argue that this is not the world we presently have.

CHAPTER 2

Issues concerning military personnel

I Professionals and non-professionals

There are professionals in the military, but the military is not composed exclusively of professionals. By professionals we mean not merely those who receive pay for work done, so that they are contrasted with amateurs. That is the sense of professional we think of when we contrast professional athletes with amateur ones. The narrower sense of professionals with which we are concerned includes receiving pay, possessing skill gained by training and includes even training that has a strong intellectual component to it.[1] Further, it includes certain organizational or institutional features. A professional in this sense belongs to an autonomous group that, in one way or another, identifies who the professionals are and then licenses them, gives them degrees, rank or some other kind of official status.[2] In effect, the organization helps set the standards so that both those in and outside the profession are better able to identify those who have reached them.

In this sense of being a professional, not all those in the military are professionals. The same can be said of many in law and medicine. Many people who help provide legal services, such as secretaries, clerks and various governmental officials, are hardly professionals the way lawyers are. Similarly, not all those who work in hospitals, clinics and doctors' offices have skills broad enough to be considered professionals. Certainly in the military, most draftees and many volunteers can hardly be thought of as professionals. Their superiors might urge them to act in a professional manner, or even tell them that they are professionals in the hope that they will act that way. But at best only some will actually deserve the label. The rest will do whatever job is assigned to them in a more or less efficient manner, all the while counting the days when they will get out of the service.

The ranks of the non-commissioned and commissioned officers will, no doubt, contain a higher percentage of professionals. If these officers are counted as professionals simply because of their official rank, their professionalism will have been achieved by definition. But if, as we are assuming, professionalism implies having certain skills and perhaps even a sense of responsibility to exercise these skills in certain ways and at certain times, it is still an open question just how many professionals are found among a military service's corps of commissioned and non-commissioned officers.

In any case, the distinction between the professionals and non-professionals is important in this and the following chapter concerned as they are with ethical issues in peacetime pertaining to personnel. So is the distinction among enlisted personnel, non-commissioned and commissioned officers, since all of the major issues discussed in this and the next chapter receive different treatment when rank is taken into account. For the sake of convenience, we will divide the discussion of these issues into three general categories: recruitment, treatment of personnel and training.

II Recruitment

The seriousness of the recruitment issue can be appreciated simply by noting that the draft is a genuine recruitment option for the military. Drafting people to serve in medicine, law, teaching or the ministry is not something we hear much about except insofar as a draft in these fields includes the military.[3] There has, for example, hardly been a ripple of discussion about drafting secretaries, let alone doctors, into medicine. The reasons for this difference between military work and that in the other professions are obvious, but need to be noted nonetheless. One is simply the numbers of people needed to serve in the military. The task of protecting a nation is often large enough to require more people than are willing to serve voluntarily. In addition, the work in the military is often strenuous, and even in peacetime it can be dangerous. Since, if war were to come, those in the military would be first to see action, military work has to be considered extremely dangerous overall. Another reason people are reluctant to enlist has to do with the changes that military life forces upon people in their life style.[4] Although those in the military may be reservists who stay home with their relatives and friends while serving, if they are full-timers they will normally render their service away from home. No doubt, some will welcome leaving home for one reason or another, but for those many who do not want to leave, a job in a local plant or office will be much more attractive. Granting that the job of serving in the military may not appeal to enough people to satisfy the needs of a country, a very special dilemma presents itself. Assuming for the moment that the needs for military personnel are genuine, either a country must pay large sums of money to encourage people to overcome their natural aversion to serving or it must resort to a draft of some sort. Putting it this way shows right off that the poorer nations of the world, if their military needs are genuine, have no choice but to draft people into the military services. Only the wealthier nations can afford to think seriously about buying as against conscripting an army, navy and air force. For the rest, they must either adopt a draft or take what consequences follow from having an inadequate military machine. These consequences include defeat in some future war, political submission to an unfriendly nation or, at the very least, status as a vassal of a more powerful but friendly nation.

As for those who can afford the option of a volunteer military establishment, they will find certain inherent problems with it. First there is cost. The up-front costs are undoubtedly higher with a volunteer military organization as against one that depends largely on the draft to solve its personnel problems.[5] It may be, as some libertarians claim, that the overall costs are less with a volunteer arrangement because of greater efficiency, professionalism and the like.[6] Yet even if this is something more than an ideological claim, which it may not be, the immediate costs of a volunteer military organization are so great as to make the volunteer option difficult even for wealthy nations.

Costs aside, there is a more serious problem for the volunteer force concept. It has already been observed that military service tends to separate its people from their home environment. The training they undergo, the on-station service they provide during peacetime and fighting they do during war are all likely to be done away from home. If the military is properly providing services to its society, it does so with a minimum amount of contact with its own people. Even when operating within the boundaries of the nation, of necessity, it does so in relative isolation. There are secrets to be kept, security problems to be dealt with, places on base that are simply too dangerous for most civilians to be around, and there is specialized training requiring isolation from the rest of society. When away in another country the isolation is more complete.

The contrast to the medical community could not be more obvious. Here the community is placed in close proximity to those who are to be served. Whatever problems the medical community has, one that it will not have is isolation from those it serves. Whenever medicine does its job properly, those who are served are there to appreciate what has been done for them. The same is true for law, teaching and most other areas in which a large number of professionals work. Although operating in a certain degree of isolation, the military establishment, at least in western countries, receives some publicity when budget-time arrives, and when called upon to perform in war or in police actions. Still, especially in peacetime, the isolation, however necessary it may be, will cause problems for the society and the military. Its citizens will not easily be able to assess how well prepared the military is to protect their interests from external and internal attacks. The only direct contact many of them will have with it will be through the taxes they pay to support it. So they will not know what the military is doing for them. They will also not know what it is doing to them. The same isolation that protects military secrets and allows the military to maintain at least some discipline over personnel, cannot help but create a vacuum so that, if doubts about its intent are raised, these doubts can easily grow unchecked.

Isolation can have deleterious effects in still other ways. To the extent that military leaders get isolated in their thinking about military and political matters, they are likely to misread what it is that the society wants from them. They may behave paternalistically, for example, in much the same way as the physician community in the United States has been accused of behaving in the past. Taking a 'we know best' attitude is the optimistic scenario here. The pessimistic one is that the military, in its isolation, would behave out-and-out selfishly by looking after its own interests rather than after those of the community as a whole.

We will have more to say about this aspect of the military's social isolation in later chapters. For now, we are concerned to talk about this matter only insofar as it impinges upon the issue of a voluntary military organization. It does so in the following way. A volunteer military service tends to fill its lower ranks with those who have come from society's lower ranks more than a conscript service.[7] Given what has already been said about the adjustments people are forced to make when they join the military, this is understandable. The hardships of military life will not seem so serious to those who are suffering certain social hardships such as unemployment, underemployment and employment in unhappy settings. To the highly employable middle and upper classes, military life with other enlisted personnel will not seem very attractive.[8] Some from these social classes may decide on a military life as commissioned officers after they have finished college. But relatively speaking, even if the middle and upper classes are over-represented among officer ranks, they will be underrepresented among the lower ranks that make up the vast majority of the military.[9] To the extent that a volunteer military force draws mainly certain classes of people it will further isolate itself from the other classes. As a result, the underrepresented classes will not be as likely to provide any direct praise or criticism of the military establishment. Instead, they will see it primarily at a distance through the eyes of the mass media, however myopic or farsighted such eyes might be. These people will not only be distanced from the military in terms of understanding what it is doing, but also in terms of emotional involvement.[10] They will not care as much as they might if they and their next of kin were a part of the peacetime military organization. If their nation were to become involved in some police action overseas, they might be concerned because, as they might put it, 'our boys are dying over there'. Yet, strictly speaking, it would not be their boys who were dying but somebody else's. The point is that this distancing of understanding and emotional involvement is unfortunate for both the people and the military. It is for the people because they will be less able to articulate what it is that they can reasonably expect from the military than they would if those in the military were more representative of the whole society. This distancing is unfortunate for the military because it will be less able to determine how best to serve the community and, in the process, sustain its own legitimate interests.

The tendency of a volunteer system to attract only certain classes of people into military life has unfortunate consequences over and above further isolating the military and large segments of society from each other. Isolation is one side of the coin. The other side is that certain classes become too closely associated with the military. Put in crudest terms, this means that the poor will bear the brunt of the casualties once a war starts.[11] Related to this charge is the claim that the poor, made up largely of minority groups, will come to resent the role they play as cannon fodder. They will sense the unfairness built into the system and, as a result, not make good fighters.[12] This argument about exploiting the poor is often coupled with another. It is said that the poor are poor in part because they are less educated. Their low level of education may not be their fault, but it is a fact of life. Given their lack of sound education, and given that, increasingly, military equipment is becoming sophisticated, a volunteer military organization systematically mismatches high-technology equipment with low-technology personnel. This leaves the military in the dilemma of suffering with this mismatch, or assuming the costly responsibility of educating the uneducated.[13] Actually, there is a sub-option available to the military under the first horn of this dilemma. The uneducated could be kept away from high-tech fields and, instead, be put in the infantry to do those jobs traditionally associated with 'grunts.' That, of course, means that this group would indeed bear the brunt of the casualties were a war to start.

This cluster of arguments having basically to do with the inequity of who serves in a volunteer military organization carries with it an assumption held by many. National defense, it is said, is not just another job the way working in business, medicine, law and education are. It is everybody's job. There is some similarity between the job of national defense and these other jobs so that it makes sense to say of a captain in the army that he has chosen his profession just as a business woman has chosen hers. No doubt to maintain a sense of continuity in the military services, as in business, there is a need for some to make life-long professional commitments to these jobs. However, so the argument goes, the defense of the nation is fraught with emergencies which necessitate that people suspend work at their normal jobs so as to help in that defense. In the past, the argument continues, this suspension of one's normal work at least for some nations (e.g., the US and Great Britain) needed to take place only during major wars.[14] Given the kinds of wars that were being fought in the past, there was time enough to collect a non-professional military so as to get the job done. But now, with a wide variety of quick-strike modern weapons available, a certain amount of military training is required before the war starts. What is needed is a large ready-reserve, and the only way to get that is through some form of peacetime conscription.[15]

We think that there is some merit to this argument. Because we do, we are not convinced by one of the major libertarian counterarguments that 'a job is a job' and that the market place ought to be allowed to sort things out by itself without governmental interference via conscription. It would make sense to agree with the libertarian if the threats to a nation came primarily from semi-organized bandits who could be suppressed by a small well-organized professional military. It would also make sense if the threats were in the distant future. But if the threats are to a nation's very existence and, again, if they are of the short-fused sort, it would seem that the market forces would work too slowly to do any good.

However, the disadvantages are not all on one side. The coercion inherent in the draft is disturbing both to rights theorists and to utilitarians. 'Yes,' the former will say, 'the rights we have to work at what we like, and travel and dress as we please put the burden of proof on those who favor the draft.'[16] 'And, yes,' the latter will add, 'those who are coerced in a draft are less likely to be happy and productive.' So there is general agreement that the draft is not good in itself. It is not only not good for those drafted, but for society as a whole because it is difficult to administer. Rarely in peacetime does a nation have to draft everybody.[17] This being so, some sort of selective service program is usually devised, one presumably that serves both the needs of society and does the job more fairly than a voluntary system. Yet the opportunities for playing favorites are obvious. Those same well-to-do individuals who under the voluntary system would not even consider enlisting now find themselves exempt from the draft. Either the laws are written in such a way as to favor them (e.g., by exempting college students) or the draft boards themselves rig things to keep the well-to-do out of the clutches of the military.

No one doubts that a draft system can be administered fairly, give or take a little. Many western nations have a draft system that seems to be avoiding gross inequities at least.[18] Yet there is no avoiding the coercion. In a military draft one's life is controlled under rules not of one's making for a period, usually, of from one to three years. As the result, for many draftees, being forced into the military seems only one notch better than receiving a prison sentence.

III Recommendations

So we find neither the voluntary nor the draft system of collecting a large number of non-professionals for the military wholly satisfactory. Neither side seems able to muster an overwhelming argument to win straight out. Instead, having the right plan seems to be a function of such things as the wealth of a nation, the threat it is living under and demographics. Concerning this last factor, it should be noted that the military-eligible group in the United States in the 1990s will be 20% smaller when compared with the 1970s.[19] Assuming that the economy will not be in constant recession during this period so that these young people really will have a choice of military as against civilian life, the military may find it difficult to attract enough volunteers to fill its ranks. As for poorer nations with unfriendly and powerful military forces at their borders (e.g. South Korea), a draft system will be the only option; while for wealthy nations with no strong enemy at their borders, a voluntary military will probably be best for them. All this is pretty obvious. What is not is what to say about these options for the next decade or so insofar as most western nations are concerned. Given what has been said already, it should be clear that the best option for one nation, even within the western camp, may not be best for another. Nonetheless, the following general comments seem appropriate. Since there are no overwhelming reasons against the draft as a way of bringing large numbers of non-professionals into the military during peacetime, this system should not be automatically eliminated as an option. There is, in fact, an argument not hitherto mentioned that suggests the draft ought to be kept as a live option. As we will argue in Chapter 8, the nuclear standoff means that nuclear weapons have no real military value.[20] These weapons cannot be used to military advantage since, if employed, they cannot be controlled. These weapons may be needed to maintain a stand-off, but aside from that, they have no other function. They enter the military calculations as expensive constants. They represent a kind of application fee that the major powers must pay in order to have the opportunity to spend still more money on very destructive, but just barely controllable, conventional weaponry. Thus if it is a conventional war for which most western powers must prepare, large numbers of trained non-professionals might be needed to fight in a future war as they did in World War II. As this fact of life soaks in, the draft may again become a live option to more nations.

The likelihood that the next major war will be a conventional one does not, however, necessarily mean that a draft is needed right away. Especially if unemployment continues to be a problem in the western countries because of a gradual transition from smoke-stack to high-technology industry, volunteer armies, navies and air forces are likely to recruit the kind of people they want with little difficulty. But there is a danger that these countries will be lulled into a false sense of security and therefore react too slowly in an emergency. The United States is probably wise then to have a draft registration system in place in case it is needed to quickly expand its ready-reserve base of non-professionals.[21]

It, and other western countries, might be wiser still if they engaged in discussions about what an effective and fair draft might look like. These discussions, we suggest, might very well yield a draft system with the following features:

1 *Universal Military Service.* This service could be thought of as a tax that everyone is expected to pay. However, instead of paying it with money, it would be paid by contributing time, energy and skills. There would be very few exemptions to this service tax. Perhaps those with very serious physical and mental illnesses would be exempt; and so might a few hardship cases.

2 *Multiple Service Options.* Those owing the national service tax would have the option of paying it by providing the society any one of several social services.[22] They could opt for military service. Indeed, the program would be designed so that it would be tempting for those in the eligible groups to satisfy the military needs of the society. Yet the eligibles could also opt to perform a wide variety of other services within their own nation or abroad. Among the services would be: (a) police work (to help control traffic, for example, so that the regular police could spend more time controlling crime); (b) social work overseas (similar to the United States' VISTA program); (c) social work at home in urban and rural areas; (d) work in education (e.g., tutoring underprivileged students); (e) public recreation work (e.g., teaching children to swim, play various games and how to keep healthy); (f) work in health care; and (g) work with the elderly (e.g., providing house maintenance services for the elderly who still live at home but can no longer do heavy labor).

3 *Service Time Flexibility.* There would be flexibility in the program in two senses. First, those in the eligible cohort would have at least some choice as to when they would serve. Normally, they would start serving at nineteen years of age. However, there would be defer-ments for those submitting a plan indicating when and how the owed service would be provided later. As an example, an engineering student might present a plan to serve in the army after completing his

degree program. Second, the total amount of service time contributed would be flexible. Those who opted for the military would serve less time than those who did not. The military draftees might serve one year and a half while the others might serve two. The flexibility in the total time served would be designed to give those in the draft a choice as to where they wanted to go. If the proper differential between military and non-military duty were found, there would be no need for a lottery system to put people in the military. There would, thus, be a choice within the system even if there would be no choice of staying out of it completely. If it turned out that certain military jobs were chosen less often (e.g., the infantry) than others, these could be assigned a shorter total service time than the other ones to make them more attractive.

4 *Reserves.* Having served the amount of time required, each draftee would be put in the ready-reserve for a period of 3 to 5 years, depending upon the needs of the country. Military reservists would be required for emergency service in their own field of regular service or any service area related to their own.

Clearly, the thrust of this proposal is to try to develop a draft system that gets certain jobs done as equitably as possible. In spite of these efforts, it may be that youths from poorer families would be lured into the infantry while their more fortunate cohorts would be attending technical schools and then serving behind the lines. Still, although such a development would represent an inequity, it is not so great as when the former group serves for two years in the military while the latter serves none at all. So the universal draft system is at least more equitable than the draft system used in the United States in the post World War II years; and it is less exploitative than the volunteer system. The equity in this system would extend to women. Given that the national service includes a wide variety of services, there would be no reason to exclude women. Even those who oppose exposing women to combat-type service could not object to giving them the opportunity to pay their service tax in other ways. Further, even those women with children could serve since they could opt to work locally in day-care centers taking care of their own and other people's children. Equity aside, a national service proposal such as this one has the advantage over a strictly military draft of mitigating the coercion inherent in any draft system. Paradoxically, it does this by coercing more people. Because the system drafts everybody, rather than a minority from the eligible cohort group, it must look for other-than-military work for the draftees. In doing so, it opens up choices for people – ones that even permit people to find work within the system compatible with their conscience.

The following three closing comments about military recruiting are appropriate. First, we are not wedded to this particular proposal. We present it simply as a proposal. Indeed, we suggest that, for the present, those western countries using a volunteer system should continue doing it. Further, we grant that our proposal seems to have a tail-wagging-the-dog feature to it. Universal service seems to be a creation to make the military draft morally viable and, to that extent, it does not appear to have much merit on its own. If that were the case, it surely should be rejected. The choice then would be between an all-voluntary military and one with a mix of volunteers and selected draftees. However, it is not obvious that drafting young people for non-military purposes cannot be justified on its own terms. The social-service tasks mentioned in our proposal are not ones that many modern societies perform terribly well. If this is so, the universal national service proposal we have made, or one like it, makes sense as a whole.

Second, it would make even more sense should the military situation for the western countries deteriorate in the next five to ten years. Given, further, the rapid technological change in the latter portion of the twentieth century and given what inevitably seems to be a shift back to conventional military strategies, it would seem to be a foolish luxury to wait to implement a draft as a war is about to start. 'No war, no draft' thinking may have made sense in the past, but it may make little sense today and even less sense tomorrow.

Third, as we have seen already, it is not obvious that a ready-reserve of well-educated and well-trained draftees would not be needed in the next war. Many projections concerning the next major (conventional) war say that it would be both nasty and short. However, these projections are not infallible. After an initial flurry of high-technology fighting, the war could continue, but tail-off in intensity so as to require a large ready-reserve. This reserve would fill in the gap between those who were needed immediately and those who could only be trained to perform some useful military function in six months or so.[23] In short, a draft would help give a nation flexibility to deal with the possibility that the short-war projections are in error as they were in 1914.

IV Treatment of personnel

Once civilians have volunteered or have been drafted into the military, questions arise about how they and their superiors should behave in relation to one another. One answer to these questions runs something like this. Upon entering military service, people lose all the rights and privileges they had before. More than that, they lose their autonomy and even their sense of identity since they are now nothing more than a uniformed part of a larger whole. When they enter the service they are like unprogrammed robots. That is unfortunate since they are useless to society in that condition. It is also unfortunate that they are not truly robots since programming robots is far easier than training humans. Another unfortunate aspect of the situation is that the tasks these inductees are being trained to perform are difficult in many ways. Complicated skills in the use of weapons need to be learned, physical strength and stamina need to be maximized, and the tasks themselves, once the war starts, often need to be performed under the worst physical and emotional conditions imaginable. To overcome these difficulties, and the further difficulty that wars are fought by groups of people who need to coordinate their activities, strong discipline is required. Unfortunately, again, such discipline in battle is not created just by asking for it. It needs to be inculcated in training over a long period of time. The training needs to be disciplined if the fighting is to be disciplined. So, from beginning to end, military personnel need to be taught to follow orders until doing so seems natural or instinctive. Less than complete obedience will not do so since what has been learned in training will be forgotten once the stress of battle takes its toll. The result will be loss of life and loss of the battle. War being what it is, and as serious as it is, requires that the discipline be brought about by corporal punishment if necessary. Harshness in training will help mitigate the more serious harshness of battle later. Truly those in subordinate positions in the military should be like robots. In the long run, if they do what they are told, and leave the responsibility for what is being done to their superiors, the job of winning battles and wars will be done more efficiently.

A variation on this total command model is different enough to deserve its own model name. Samurai in Japan were, by the meaning of that term, warriors who serve. They might be in service to the emperor, a daimyo (provincial lord) or a shogun (a great general).

Although samurai were not always expected to express agreement with their lord if they thought that what they were commanded to do was foolish or dishonorable, their code of ethics put great stress upon obedience. Even if, in serving the lord, death were almost certain, the samurai were expected to face death honorably. The following brief passages from the *Hagakure* give a sense of how the samurai were to face death and of their relationship to their superior.[24]

> The Way of the Samurai is found in death. When it comes to either/or, there is only the quick choice of death. It is not particularly difficult. Be determined to advance. To say that dying without reaching one's aim is to die a dog's death is the frivolous way of sophisticates. When pressed with the choice of life or death, it is not necessary to gain one's aim.[25]
>
> Being a retainer is nothing other than being a supporter of one's lord, entrusting matters of good and evil to him, and renouncing self-interest. If there are but two or three men of this type, the fief will be secure.[26]

Insofar as samurai are obedient to their masters, even in the face of death, they and the society which bred them seem to fall neatly into the total command model. Even in disagreement, they seem to do so since it is not an option for them to leave their lord's service saying something like 'Our consciences do not allow us to participate in the activity you have mandated.'[27] Instead, they often expressed the sincerity of their disagreements by committing seppuku (hara-kiri). There were other reasons for committing seppuku, such as to avoid capture and as atonement for dishonor. But the way the samurai expressed their disagreements showed that they did not take their lord's commands lightly.

This appearance of acting within the total command model is a bit misleading, nonetheless. It is true that in some samurai traditions, there is the aspect of mindless obedience to their master. However, not all samurai should be thought of as robots. To some extent, at least, they had an ethics of their own that involved disciplining themselves rather than having the discipline imposed on them by their superiors.[28] The orders might come from on high, but the virtues of the samurai were largely internal to the men themselves. No doubt, something like this same sense of self-imposed obedience was found among the medieval knights of Europe. Still, given this difference between the samurai and the knights, on the one side, and the soldiers of the command model, on the other, it is best to treat the former model as a distinct variant of the total command model.

One other difference between the samurai and the non-professional inductees found in the total command model is the image of the former as 'hired guns'. Although some samurai were born into warrior houses, others looked for employment. In this sense, this latter group resembled those operating under what we will call the mercenary model. Mercenaries are, in effect, samurais without a sense of loyalty. Like the samurai they are professionals in the narrow sense of that concept of having mastered certain skills related to fighting, and in the sense of sustaining themselves through the exercise of those skills. But, unlike the samurai, the loyalty that mercenaries have to their employers extends only as far as it makes good business sense to do so.

It was this limited sense of loyalty that loomed heavy in some minds in the early 1970s when they opposed the notion of an all-voluntary military organization in the United States. The so called Gates Commission Report, which set the framework for the all-volunteer force, emphasized the notion of serving in the military as an occupation.[29] Many thus express the fear that the enlisted ranks in an all-volunteer force would soon be made up of mercenaries. In point of fact this fear was misplaced. It was not as if the enlisted volunteers in the US military were likely to offer their services, en masse, to the Russians, Argentinians or Saudis should they be offered better salaries and working conditions. It was not their loyalty that was in question, as the label 'mercenary' implies, but their motivation. The question was whether those military personnel recruited mainly because of money would work as hard and fight as effectively as those who were drafted. It is not totally clear what the answer to that question is. What is clear, however, is that if the US military were a mercenary organization in the more proper sense of that term, it would not fill its ranks primarily with Americans, but would find personnel from client states who would be willing to sell their military skills for a lower wage.

Two other models of military organization need to be mentioned. It is at least conceivable that a totally non-authoritarian military organization could be established. Presumably decisions made within such an organization would not be made in accordance with rank. Instead, each member of the group would vote in order to help arrive at a consensus as to what should be done. Such an arrangement might come to some reasonable decisions during peacetime but, during battle, making decisions would surely be more difficult. Presumably, in this latter setting, one of the soldiers would have to be designated to implement the will of the group by announcing what each battle group is expected to do and where and when it is to do it. We mention the totally non-authoritarian model not because it represents a viable option. Rather, we mention it because it helps in characterizing still another model – one that falls someplace between it and the total authoritarian/samurai models. Morris Janowitz talks about it in *The Professional Soldier*. He argues that the total authority model is no longer viable at least for western countries. Instead, he sees that a shift has taken place from an authority or command model to a leadership model that emphasizes persuasion rather than coercion.

The shift from domination to manipulation and persuasion involves the relative balance of negative sanctions versus positive incentives. Domination is defined as issuing orders without explaining the goals sought or the purposes involved. This was the spirit of the charge of the British Light Brigade. It came to an end only after the battles of the Somme and Paaschendaele, when Allied civilian leadership began to see the pyrrhic victory such actions would bring. Manipulation implies ordering and influencing human behaviour by emphasizing group goals and by using indirect techniques of control. While the terms manipulation and persuasion have come to be thought of as morally reprehensible, they describe the efforts of organizational management when orders and commands are issued and the reasons for them are given. It is impossible to analyze modern bureaucratic institutions without reference to a concept such as manipulation or persuasion, or some more socially acceptable equivalent. The objective of the effective military manager is not to eliminate differences in rank and authority. Instead, he seeks to maximize participation in implementing decisions at all levels by taking into consideration the technical skills and interpersonal needs of all concerned.[30]

Janowitz gives several reasons why this shift has taken place. One is purely military. Close formation fighting which required 'direct and rigid discipline' no longer makes sense given the powerful and rapid firing weapons available today. Looser formations dictate that the soldier must learn to fight more on his own.[31] A second reason is partly military, partly sociological. The technology of warfare is so complex that the coordination of a group of specialists cannot be guaranteed simply by authoritarian discipline. Members of a military group recognize their greater mutual dependence on the technical proficiency of their team members, rather than on the formal authority structure.[32] Evidently, the skills needed in a military organization are too diverse for anyone to lord his rank over others in an all-knowing manner.

A third reason Janowitz cites for the shift away from the authoritarian to a leadership model is purely cultural. Times have changed. A military organization within a society cannot operate in complete isolation from the cultural changes within that society.[33] Businesses, schools, churches and families are all less authoritarian than they were at the beginning of the century. As they have changed, so has the military. It is simply not possible for the military to take large numbers of men and women and transform them into followers of authority when they have been raised in a tolerant society.

Janowitz's reasons are primarily causal, not ethical. He is telling us why certain things have happened, while our concern is with how they ought to happen. What is happening may not be the same as what ought to, but in this case the *is* and the *ought* happily coincide. If coercion ought to be kept to a minimum, then we ought to be pleased that technology has schemed to make coercion less necessary than before when it comes to training and fighting. We ought to be pleased also that technology forces those in control of the military to seek the participation of a wider segment of the population in the decision-making process. This is not to say that the leadership model is or should be changing gradually until it becomes transformed into the non-authoritarian model. Rather, the picture of what seems appropriate can be characterized as follows.

Under the leadership model, military personnel will ideally possess four general features:
1 They will have skill in the use of their weapons.
2 They will be disciplined in working with others and be able to improvise when left to fend for themselves.
3 They will be motivated.
4 They will have a sense of moral responsibility for their and their group's actions.

The first of these four features almost mandates that at least weapons instructors (mostly non-commissioned officers) have some authority. As we have already noted, weapons today are far more complex than in the past. Teaching the skills to use them takes time and requires that those with the knowhow have the authority to say and do the things necessary to help others gain these skills. The inappropriateness of the non-authoritarian model is evident here since high degrees of strictness in teaching how weapons should be used are required for both effectiveness against a potential enemy and for the safety of the users. Recruits may not fully appreciate the dangers associated with using weapons even when explanations are given. So it is inevitable that the leader under this model will be, at times at least, a coercer.

The second feature should be divided into two parts, with the first requiring coercion as well. Quite apart from the skills needed to use weapons and other military equipment, there is the problem of coordinating their use with others and doing so under less than ideal conditions. So superiors, using their authority, are required to teach and maintain discipline. We will have more to say about this in the next chapter when we discuss the process of teaching codes of ethics to the non-professionals.

The second part of this second feature appears to be in contradiction with the first. However, it is not. There is no psychological reason why people cannot, at certain times, respond consistently to external discipline and, at other times, exercise the abilities needed to act on their own. To think otherwise is a mistake of the total authoritarian model. It may be true that if battles were always and entirely fought with superiors standing at their inferiors' shoulder, the best fighting groups would be those operating under the strictest standards of the total authoritarian model. Since, however, battles are often not fought that way, it becomes imperative to compromise by teaching military personnel when and how to follow orders and when and how to act on their own.

Whereas discipline is achieved by means of the liberal application of authority in the use of weapons and in coordinating their use with others in military organizations, motivating military personnel requires more varied personnel policies. No doubt, on some occasions, a high level of motivation can be achieved and maintained within the framework of an authoritarian learning environment. Pride in surviving the strains of a disciplined training program can do wonders for morale. Rewards and promotions can further enhance military morale. However, it is not likely that these considerations will sustain high morale and motivation. As Janowitz noted with US military personnel, and this point is applicable very likely to most western nations, if the society is non-authoritarian it is not likely that those drafted or volunteering into the military will thrive on an exclusive diet of authority. If people are accustomed to making their own decisions, they will be happier and work more efficiently if the authority the military imposes on them is as circumscribed as possible. We must keep in mind that all western nations who maintain a strong military posture must depend on non-professional military personnel to fill out their ranks. These non-professionals among the enlisted ranks might be draftees or they might be volunteers who find themselves in the military for any number of reasons. Some of the latter group might, as we have seen, view the military as the employer of last resort. For people like these, and for those who are reluctantly in the military as the result of a draft, something other than total authority will likely be required to help them maintain a high level of motivation. Minimally what will be required is a periodic accounting of why these, rather than some other, orders have been issued. Neither blind obedience of the total authoritarian model nor the samurai model seem applicable to the many people who find themselves in military camps for reasons other than their love of military life.

In order to transform the military from anything like the total authoritarian to a leadership model, 'why questions', having to do with more than just orders, will probably also need to be answered. To maintain a high level of motivation among people raised on a diet of non-authority, explanations will be needed about more general matters such as why it is important for those in the military to be where they are, and even why it is important for them to be in the military in the first place. Unfortunately, establishing procedures for answering these questions raises some very difficult problems for the military and, indirectly, for the society as a whole. In effect, what the military is being asked to do in dealing with these 'why questions' is educate its personnel about social and political issues. This is not likely to be an assignment that many military leaders will relish if for no other reason than that such education would take time and resources away from teaching purely military tasks. But even if others in the military would be willing to take on this task, it is not clear that enough trained people could be found to do a proper job. Then there is the further danger that the social and political lessons would be presented in a self-serving way. The educational process might just turn out to be more indoctrination than anything else. If it were, it would probably be received negatively by the students and at best constitute a waste of time and money. Yet even if it were effective, but in a misguided way, it might do more harm than good. Using their station as military leaders, those in charge of motivating their subordinates could very well present misinformation and biased viewpoints that would make those listening motivated to serve, to be sure, but to serve the military rather than the society.

Nonetheless, with all the dangers inherent in this process of educating military personnel about social and political issues, we will argue that they are worth risking. As we have suggested, the issue is not how to motivate those in the military so that when the time comes they will fight and fight well. A variety of social strategies might bring that about, including threats, propaganda and bribes. Our concern here is motivating military personnel to fight both well and morally (the fourth feature of the leadership model). We are suggesting that people will fight better if their cause is just and they believe in it. Put this way, it seems almost absurdly obvious that a society should educate all of its military personnel as to its virtues. Nonetheless, as we will see shortly, objections to such an educational program will not go away easily and, further, are complicated by the close connection the fourth feature of the leadership model has with the third feature (i.e., motivation). It will be recalled that the fourth feature is that military personnel functioning within that model should have a sense of moral responsibility for their and their military's actions. Clearly the connection between the third and the fourth feature is that they both require an educative process. Eighteen-year-olds moving into the military do not necessarily have a well-developed sense of moral responsibility. And even if they do, it is likely blunted by the heavy peer pressure of the barracks. Given, then, that educative processes concerned with the third feature (i.e. motivation) and the fourth one (i.e. morality) are so closely related, we will make recommendations about them side by side as we focus in the next chapter on moral education.

CHAPTER 3
The place of codes of ethics in the military

I Sceptics and idealists

Eventually we will argue that preparing military personnel to deal with moral issues is no easy task. To get it done right, a variety of educational and disciplinary strategies need to be implemented. However, since probably the most important part of that task is to teach military personnel certain codes of ethics, we will start by discussing these codes. And pertaining to codes, we will first deal with what place or places they have in the military, and then later with what they might be like. After that we will go on to describe the other aspects of the process of preparing military personnel to actually deal with moral issues. Finally, we will describe how this preparation ties into the political/educative process that we said in Chapter 2 is necessary if a military organization is to function well (i.e., both morally and militarily) within a community.

The reason it is important to start the discussion of codes of ethics with the place they have within the military is that their proper place there is more often than not misunderstood both by sceptics who attack such codes and by idealists who defend them vigorously. Those we will call sceptics do not necessarily deny that ethics has a place in military settings. What our sceptics are sceptical about are the codes themselves; and they are sceptical because they misplace or misconceive what these codes are supposed to do. Likewise those we are calling idealists concerning these codes misplace them by coming to expect too much from them. So understanding the right place for these codes is no trivial academic exercise. It makes a difference where the codes of military ethics are placed as to just how they are used and misused.

An initial step toward understanding the place of codes of military ethics can be taken by referring once again to R.M. Hare's distinction between the critical and the intuitive levels of thinking.[1] It will be recalled that critical thinking about moral issues questions the rules and principles on the intuitive level and deals with conflicts between the rules and principles on that level. Because the critical level examines the intuitive, but not vice versa, the intuitive level can be said to be parasitic on the critical.

However, this asymmetrical relationship between the two levels does not mean that the intuitive level is unimportant or dispensable. For one thing, this level of thinking is needed when raising children. They cannot be expected to think critically in their early years; yet they must learn in these years to behave in accordance with society's ethical norms. Exposing them to some arbitrary ready-made rules which, in time, will seem intuitively true will get them started on the road to behaving properly. For another, and this is more important for our purposes, there is simply no time for critical thinking during certain emergencies. Time is not always available to critically think about whether it is morally proper to torture a prisoner in order to extract information from him about where, beneath us, he has planted a bomb that is about to go off. At such a point in time, all we can do is appeal to our intuitive or ready-made rules. Hopefully, the rules we have, if we have any at all, have been critically examined prior to such an emergency.

According to Hare, the place of codes of ethics of all kinds is on the intuitive level. Codes are paradigm examples of sets of ready-made rules. In addition to being useful in educating children (or raw recruits in military settings) and during emergencies, they save time by keeping us from having to think through what we are supposed to do on every occasion when we are faced with a problem. Without such rules, we would be kept so busy doing critical thinking as to be paralyzed into inaction.

II Comparison to medicine, law and other fields

Having placed codes of ethics, specifically military ones, on the intuitive level gives us only a general sense of the place these codes have within the military. This is because military codes have some special features that they do not share with codes in other fields such as medicine, law, education and business. It is useful, therefore, to make comparisons with those other fields to find more precisely what the proper place is for military codes of ethics. One way to make comparisons is with respect to the content of these codes. As one might expect there is some overlap here. All the kinds of codes we are comparing express a concern for the integrity of their respective fields or professions, and all take the high road in speaking of service, honor, honesty and loyalty. Further, the wording of these codes suggests that they are concerned with very weighty matters. Yet one cannot help but be struck by how much more important some codes are than others. For instance, military codes have more to do with life and death matters than codes for education, law or even medicine do. There is a kind of ethics of scale involved here. Whereas physicians may have to make decisions of life and death about individual patients, military people routinely make decisions that affect the lives of scores, hundreds, or even thousands of people during war. Further, they make these decisions not just as they affect people who are marginally alive because of some illness but, more often than not, as they affect those who are young and in good health. A plausible case, therefore, can be made for saying that codes of ethics in the military have at least as much weight or substance to them as do codes in other fields.

Military codes of ethics differ from other codes in ways other than content. Codes are not just lists of rules. They are lists intended to apply to certain people. Thus within the medical community there are codes for physicians, nurses, and other health-care specialists. In medicine's case, these codes center on the physician as can be seen in the nurses' codes which stress, among other things, obedience to the physicians. Obedience of inferiors to superiors is found in military codes also, but with an important difference. In medicine the significant actions that could have the greatest ethical implications are performed by the physicians. They are the ones who examine their patients, perform surgery on them and literally treat them on a 'hands-on' basis. They also give orders concerning routine matters to others. Military commanders certainly give orders in a similar fashion, but beyond that they do not, strange to say, do very much. Above a certain rank, commanders will not regularly be found on the field of battle shooting at the enemy. Instead, they become administrators who have others do the shooting for them. In medicine that would be comparable to physicians devising plans for surgical procedures and then ordering the nurses, other health-care professionals and non-professionals to actually carry them out.

This comparison to medicine helps to show that the range of

assigned responsibility in military matters is extremely wide since it is those in the lower ranks who, as it were, do the surgery for the generals and admirals on the enemy. This means that in contrast to practices in medicine, and even law and education, *both* professionals and non-professionals in the military do many things that inevitably have major ethical implications.

The necessity of war forces this sharing of assigned responsibility across all ranks, where those with high rank take responsibility for ordering and those with low rank for actually doing whatever needs to be done. The problems these arrangements of responsibility pose for us are twofold. First, from the point of view of those who give orders, what gets done becomes remote, so it becomes difficult for them to assess militarily and morally what is happening. About such remoteness A.J.P. Taylor in his *The History of the First World War* gives the following account:

> Paaschendaele was the last battle in the old style, though no one knew this at that time. Even the generals at last realized that something had gone wrong. On 8 November (1917) Haig's Chief-of-Staff visited the fighting zone for the first time. As his car struggled through the mud, he burst into tears, and cried: 'Good God, did we really send men to fight in that?' His companion replied: 'It's worse further up.' Haig alone was undismayed. He went on planning a renewal of the campaign in the spring.[2]

It may be true that military commanders at all levels are encouraged by military doctrine not to lose touch with reality as presumably Haig did. But as we all know, the chain of command is very long, thin, and sometimes broken, and thus the truth about what is going on in battle often gets lost long before it reaches headquarters.

Second, from the point of view of those doing the shooting, their sense of moral responsibility will often be no match for the levels of responsibility that are laid at their feet by their commanders and society. As many of them will be non-professionals, they will likely have a less fully developed sense of what is morally expected of them than will those professionals who are commanding them but not doing the fighting.

Thus, the special place codes of military ethics have is that they apply both to those who are and are not well trained to follow their edicts. As if this were not bad enough, the edicts apply to those doing the fighting under the most trying conditions. These fighters are expected to live up to certain standards of ethics while under fire, while suffering from fatigue and injury and while fighting under less than ideal environmental conditions. It seems paradoxical that we expect them in battle to deal with the most serious moral problems we can envision with little or no professional training and do so often while working under the worst possible conditions.

Thus far, everything seems to be more serious and difficult when applying codes of ethics to the military as against other fields. Taking account now of another consideration which makes the military setting different from the settings of health care, law, and education seems, if anything, to make matters worse. This consideration has to do with the special group character of military activity. Both sociologists and all those who have been in a military uniform are familiar with the kinds of group behavior exhibited by young men in uniform. Such behavior can, and at times does, range on the negative side all the way from mild intimidation of others to rape and murder. In a group, these men will do things they ought not, which they would not do by themselves. The principle here seems to be that the whole is worse than the sum of the parts. This same kind of behavior becomes still worse when the group is given weapons, transported to another nation where the social constraints upon it are fewer than at home, and allowed to operate within a permissive military command structure.[3] When this, along with the other group characteristics of the military discussed already, is taken into account, it becomes difficult to avoid the conclusions that greater moral constraint is required within the military than in medicine, law, education, etc., and that if codes of ethics can help to bring this constraint about, then they are needed more in the military than in these other fields.

If this greater need is indeed present, little as yet has been said as to how it might be satisfied. Fortunately, the answer to that question is found in still another difference between the military and the other fields with which we have been comparing it. Comparisons to medicine and law are again particularly useful. Institutional organization in these fields is much looser than in the military. Some organizational discipline is certainly present during the training periods in medicine and law, although the control is far less than it is in basic training within the military. Once training is over, a certain amount of organizational discipline will continue for those in medicine who work within the framework of a hospital or medical institute. However, many others will practice their profession as physicians (or lawyers) as Lone Rangers. No doubt those professionals who do their work mostly by themselves will remain subject to the constraints of their professional associations. Still, these associations are often permissive in practice, if not in principle. So if they have a code of ethics, the members will be expected to impose the code upon themselves. The codes will be seen largely as formal-like appeals to conscience, self-respect, and professional integrity. As a result, some will be strongly tempted to view these codes in particular and any call to ethical behavior in general as just so much mouthwash.

In the military, codes of ethics can also be seen as gyroscopes to guide individual behavior. However, since the military viewed as an institution is much more authoritarian than medicine and law are, it need not accept the idea that codes of ethics are merely for individuals to adopt or not adopt as they see fit. Just considering this authority by itself, great success might be expected in implementing codes of ethics. However, as we have observed already, the ethical problems of the military are more serious and difficult than in other professional fields, so it is not clear how much success can be expected in the end. The military's greater authority and its greater problems may just balance each other out, leaving the military with codes no stronger in principle than those found in medicine, law, business and education.

III The sceptics' arguments

Nevertheless, the strong need for codes of ethics within military settings encourages persistence in the inquiry into the possibility of having functioning codes. It will be recalled that the sceptics have serious doubts that codes of ethics have any place in the military scheme of things. We did not specify why these sceptics have these doubts, but it should now be clear that these doubts probably rest on their awareness, dim or clear as it may be, of the difficulties facing those who would put the brakes of ethics on a military machine. Others might be sceptical simply because they are realists about war who feel that ethics, whether in the form of codes or not, does not belong in military settings. Yet still others might be sceptical for a totally different reason. These people may feel that ethics has a place within the military, but that codes themselves are a waste of time for reasons other than those mentioned until now. Some of these people might feel that these codes have a tendency to be held too rigidly once they have been accepted, while others in this group might feel that codes in large enough numbers have already been put down on paper. Are more codes needed, this latter group wonders, so they too can be ignored?

As we have already indicated, we want to argue against the sceptical attack, if not necessarily against all the range of reasons which support it. We must, of necessity, oppose the realist reason some sceptics give; otherwise the whole study of military ethics becomes moot. We need not, however, disagree with those sceptics who attend to those conditions that militate against having an effective code of ethics. Indeed we have insisted that these difficulties are real. Further, we need not even disagree with the sceptics who say that there are too many codes to be found within the environs of the military. In fact, most standing military organizations have had ample opportunity to codify and recodify the ethics with which they are most comfortable. It remains possible that whatever failure they are experiencing in getting their military personnel to behave morally is due to the need for further recodification. They might not yet have gotten it quite right. Still, the felt need for codes, even though they are already in existence, may not be really one for new or modified codes as such. Rather, the need may have to do more with effectively promulgating the codes already in existence.

IV Promulgating codes

If this suggestion has merit, it is important to consider how codes can be promulgated in more detail than we have up until now. We will begin this consideration on the lowest level of the military ladder. We have said that the non-professionals in the military pose special problems. Being less trained in military traditions than the professionals, and less educated generally than their leaders, they will have more of a tendency either not to know what to do when faced with an ethical problem or to do the wrong thing instinctively. No doubt many will bring a sense of right and wrong with them from their religious and educational background. Yet many will not. Once the mix of those with and those without this sense takes shape, only the most optimistic will assume that right will dominate wrong. Reminding ourselves also that this mix of non-professionals could very well be asked to act militarily under the most stressful conditions, it would seem that promulgating the codes merely by giving a lecture or two in basic training would hardly be enough to do the job. Codes, whatever their content, would have to be so ingrained into the consciences of military personnel as to be unforgettable. As to those without a conscience, it would be enough for them to ingrain these codes into their general memory with the added thought that violations would be met by sanctions from those above them in rank.

The exact form the training in the codes would take would vary with the circumstances. If a war were underway or about to begin, and the time the non-professionals were in the military was counted by the weeks, promulgation might mean little more than committing the codes to memory by repeating them in training sessions and, in addition, making clear what the procedures are for invoking the codes against those who act immorally. This suggests that the codes would have to be brief in form and to the point. Prisoners, it might say flatly, are not to be killed, tortured, robbed or abused. In the rush to battle there might not be time to deal with such borderline problems as when an enemy soldier is a prisoner. Is he one simply by dropping his weapons and throwing his hands up? What if he does these things in the heat of battle when there still is some danger involved in the process of taking prisoners? The promulgation of the codes here would not include such refinements in the art of ethics. Rather, it would be designed to help the military avoid committing gross moral errors such as those at My Lai in Vietnam.

In peacetime, elaboration on the codes would be possible. These codes would still need to be kept simple in that each one would likely need to be kept to eight or ten rules with each rule kept as simple as possible. Within that format, the content of the rules would have to be specific enough to have prescriptive power, and the major content areas would have to be encompassed. Certainly a rule about obedience to a legal order from a superior would be included, as would one concerned with truth-telling and the treatment of civilians. But now in peacetime, while there is time, elaboration would take account of more exceptions as well as borderline problems. Explanations about codes in general could also be encouraged such as, for example, that they are revisable and that they are not intended to solve all the moral problems we encounter. These elaborations and explanations could be presented to the lower echelon personnel with case studies in a discussion format so that they might develop a sense that the ethical problems are real ones. In all this educative process, however, it would be a mistake to lose sight of the basic rules and to allow those who are to follow them to forget what they are. Part of the process of learning about the codes would be that deviations from them are at most very rare and that therefore when in doubt it is probably better to follow the code than not.

In sum, codes for the lower military echelon will be made up of concrete rules each of which is to be followed rigidly and automatically. They will be viewed as codes that have the sanction of the military and the society which stands behind them. Such codes, therefore, are not to be violated lightly.

To some extent the place of codes of ethics for NCOs will be much the same. They, too, will have to memorize them rigidly and keep in mind that penalties will come to those who are forgetful. However, these professionals will be giving orders as well as taking them, so the codes they follow will be somewhat more complicated. They will also function as teachers of the codes, so their understanding of them will have to go beyond that of those below them. In fact, they may be more important as teachers to the enlisted ranks than are the commissioned officers. After all, they spend more time with their 'students' so they have ample time to teach the codes and other things both by example and in lesson form. Additionally, NCOs are not so distant from those under them in rank; and at least the good ones can, as a result, relate to them better than can the brass. It would, therefore, be a great mistake to rely exclusively on commissioned officers to promulgate a code of ethics.[4] Just as in most things, what the enlisted personnel learn come from the NCOs, so the codes must come to a large extent from the same source.

To some this conclusion may seem unfortunate, since a sergeant or a petty officer may not seem to be the right sort of teacher from whom to learn about ethics or morals. It might seem more appropriate to have the chaplain do that job if it is true that the regular commissioned officers need some help in this regard. But ethical rules put together into codes are basically no different from rules of any sort that military people have to follow. If a sergeant can teach recruits the rules about how to behave on the firing range, he can teach them about obedience, truth-telling and the treatment of prisoners. This is not to say that the NCOs will do all of the teaching. Presumably they will be supported by commissioned officers who will, in addition to teaching enlisted personnel, be responsible for teaching the code to the NCOs and to other commissioned officers. The officers, the NCOs and even some enlisted personnel have still another task with respect to whatever codes are developed. It will be recalled that, according to Hare, codes belong on the intuitive level of moral thinking. This is the level where the rules are learned so well that they are followed automatically. In this regard, what we have said is that codes are to be drilled into military personnel in the same way that other rules are. But there is also work to be done on the critical level. Here certain officers, and others, need to review them to see if they need revising. This suggestion merely reiterates the point made earlier that codes are not final statements of some truth. Those who treat them as such and then refuse to accept emendations or exceptions to them represent the idealists mentioned at the beginning of this chapter. To say that critical thinking will from time to time make changes in the codes is part of what it means to say that codes should not be taken too seriously. Thus, officers need to follow those rules in the codes that everyone else follows. They need to follow rules that apply to them as officers; they need to be involved in teaching the code; and now, it appears, some of them also need to do critical thinking about the codes. More on all this shortly.

We have spoken of the danger of attack from both the sceptics and the idealists. Against the sceptics we have argued that the seriousness of war and the group behaviour of young armed men create a strong need for rigid codes of ethics. We have also argued that the tight institutional structure of a typical military organization makes it possible to promulgate such codes. Against the idealists we are now arguing that although codes should be rigidly held to, especially in emergencies, they should not be taken too seriously in two senses. First, they should not be thought of as the whole of what military ethics is about. Codes merely guide us in those serious situations that occur repeatedly enough to warrant our attention. It is difficult, and very likely impossible, for codes to cover all contingencies without becoming so abstract as to lose their prescriptive character. Second, codes are not fixed in concrete. Like other 'laws' they are subject to emendation by critical thinking. So whatever code is in force, it is not the all and forever of military ethics.

What, then, can be said about the place of codes of ethics within the military up to this point? Codes are practical devices for controlling people's behavior in recurring situations. The idealists' mistake is to make these devices impractical. They do this by creating codes with standards so high that few, if any, can reach them. In their idealism, they also create the impression that codes are immutable and that they somehow blanket all possible moral contingencies. But codes are neither immutable nor all-encompassing. Critical thinking enables us to revise codes as situations change. And the very fact that codes cover only recurring situations shows they cannot be all-encompassing. They further cannot be all-encompassing since codes need to be brief. Brevity is especially necessary for military codes of ethics, since one of their primary aims is to help control the behavior of non-professionals who have less sense of military traditions than professionals do.

It is also important to know how to use codes as practical devices. It is in this regard that many sceptics misplace military codes of ethics to the point of supposing that they have no place in military life at all. It is no good just telling people about the codes they are expected to live by. These codes must be properly explained and inculcated, and rigidly enforced. Once they are, they need no longer be seen sceptically as a waste of time and effort.

V Types of military codes

It is not our intent now, or even later, actually to present a, let alone *the*, code of military ethics. Presenting such a code is inappropriate since it would best be developed by a panel of people, some who are specialists in ethics, law, and like fields, and others specialists in various aspects of military activity. What we will do, however, is outline in more detail than we have already, what a military code might look like if properly done. And, as we have suggested already, the first thing to note is that it would not be just a single code. All along, the discussion has been in terms of codes of ethics not in the sense of devising alternative codes to do one job, but devising more than one code in order to get more than one job done.

Thus far the main focus of attention has been on a minimal code of rules to guide the behavior of professionals and non-professionals, especially the latter, in battle situations. Such a code would be different from one aimed at guiding the behavior of military personnel when they are in captivity. In turn, there would be differences between these two codes and one whose purpose is to articulate the proper relationships that should exist among military personnel – both equals and unequals – within one military organization. There might also be a rather different kind of code from all of these since it would speak more to the ideals of the military profession than to the minimal or basic kinds of behavior that should guide military personnel. This so-called creedal code would not be totally different from the others since these latter codes might also make reference to ideals. However, it would be different from them since as a goals-oriented or aspiration code, failure to live up to it would not result in punishment normally. It would be otherwise with the other codes. If, for instance, an officer were discovered to have tortured and sexually abused a prisoner, swift retribution should follow. But if an officer failed to achieve the lofty standards of a creedal code that speaks of honor and bravery, he might not be promoted or he might be shunned, but normally he would not be punished.

To be sure, there is a danger of a proliferation of codes. But four, possibly five, ought not to be too many. Certainly one is too few. Trying to encapsulate meaningful items in a code that is brief and yet encompasses all the major aspects of military life is too much for any one code to bear. In any case, there is no real need to get it all down into one code except that people traditionally have talked about a code of ethics and then, without paying much attention to how they are talking, have slipped into expressing themselves in terms of *the* code. However, since military life quite naturally breaks itself down into a few basic relationships, a code for each relationship can quite easily be developed.

The *internal code* would govern relationships between personnel within the same military organization. This code would be especially important for a military organization during peacetime, since it is during these long periods before wars begin that such a code gets implemented in a real sense. It is during peacetime, for example, that physical abuse by superiors of inferiors is either kept at a minimum or allowed to flourish. Very likely what happens in the relationships of military personnel during war is an extension of habits and traditions established during peacetime. Further, such a code would be the first one taught to those entering the service since the other codes presuppose it. After all, meeting and establishing relationships with one's own people are things that happen long before the enemy is seen even at a distance.

As to the code itself, it should contain rules governing the basic military relationships existing between those in any military organization. Items pertaining to the following areas would certainly be

included in this code:
1   following orders (legal ones only, presumably);
2   giving orders (e.g., that are not primarily aimed at profiting the order-giver);
3   abusing others (e.g., not allowing a superior officer take sexual advantage of an enlistee);
4   treating people fairly (e.g., not permitting a sergeant to give preferential treatment to his buddies when unpleasant work has to be assigned);
5   doing one's job conscientiously (and thereby helping the community – both the military and the society as a whole).

It makes sense to introduce *the creedal code* along with the internal code. Indeed, the creedal code could be presented as a kind of preamble to the internal code and the fighting code (discussed below). Whether separate or not, the creedal code would take one of two forms. It could be inspirational by speaking to such general concepts as duty, honor, country. If it took this form, the assumption would be that the pivotal terms and phrases mentioned in it would gain their prescriptive bite not solely from whatever linguistic meaning they carry but from that meaning plus the context (i.e., the military tradition) in which they are embedded.

The other form the creedal code could take would be a list of the virtues of the ideal man (person) of arms. Even this form of the creedal code would seem general when compared to the rule-oriented internal and fighting codes. Yet, rather than being inspirational, it would explain, one at a time, and in greater detail than the inspirational form, the relevant military virtues such as obedience, bravery, honesty, fairness, steadfastness and leadership. No doubt, in either form, the creedal code would have its greatest appeal among the officer classes who, for a variety of reasons, would have a better sense of what the general concepts contained in that code mean.

The *fighting code* would look very much like the internal code in placing emphasis on rules rather than virtues, and would also overlap it in content. It would, for instance, very likely make reference to following orders, but its main thrust would be to make 'external' prescriptions. Primarily it would speak to how the army, navy, and air forces should behave towards the enemy military personnel and to all civilians. We have said much already in the early portion of this chapter about this particular code. Suffice it, for now, to add that especially when non-military people think of military codes, this is the one they have most in mind. And understandably so. If we take seriously the notion of morality (or more properly immorality) of scale, it is when military personnel are using their weapons that the greatest abuses of moral principles are likely to occur.

The final code of ethics has to do with how one should behave when in danger of being, and after one has been, taken prisoner. This *prisoner's code* of conduct is needed especially when the enemy treats prisoners in an immoral fashion by means of isolation, torture, starvation and use of tactics that encourage prisoners to become alienated from their fellow prisoners. This code would cover conditions under which it is permissible for someone to surrender, what one should do about escape, what kind of information one should give to the enemy, how one should cooperate with the enemy, and how one should respond to fellow prisoners and to senior ranking officers.

It may be that some would prefer to have three instead of four codes, perhaps dispensing with the creedal code. Others might think that another code or two is needed. The exact number of codes is not what is at issue. What is, is that having just one code is too few and ten is too many since, once again, one code has to bear too high a burden, while ten would make things too confusing. What is also important is that the three to five codes should be kept distinct from one another. Each, after all, is concerned to speak to radically different settings in which military personnel find themselves. Further, in creating just a single code, there is a strong temptation to load it up with rules of one kind to the neglect of another. For example, a code created by a group of officers might be inspirational almost exclusively and thus not speak to certain internal and external matters.

VI Critical thinking in the military

Having and promulgating four (plus or minus one) codes is not a substitute for doing other things to bring about ethical behavior among military personnel. It is not as if codes (representing intuitive thinking) must be chosen over an educative process emphasizing critical thinking. Intuitive and critical thinking are fully compatible here. Of course, promulgating the codes is itself an educative process. The codes, as we have seen, are not just to be recited in some mindless manner but also to be explained. In this connection, it is interesting how the US Army manual 21–78 titled *Prisoner of War Resistance* not only presents six articles in a code very much like what we are calling a prisoner's code, but also presents an explanation of each article in that code. Article V, typical of the others, reads as follows:

> When questioned, should I become a prisoner of war, I am required to give name, rank, service number, and date of birth. I will evade answering further questions to the utmost of my ability. I will make no oral or written statement disloyal to my country and its allies or harmful to their cause.[5]

The explanation begins directly following the statement of this article.

> When questioned, a prisoner of war is required by the Geneva Conventions, this Code, and is permitted by the UCMJ to give name, rank, service number, and date of birth. Under the Geneva Conventions, the enemy has no right to try to force a USPW to provide any additional information. However, it is unrealistic to expect a PW to remain confined for years reciting only name, rank, identification number and date of birth. There are many PW camp situations in which certain types of conversation with the enemy are permitted. For example, a PW is allowed but not required by this Code, the UCMJ, or the Geneva Conventions to fill out a Geneva Conventions 'capture card', to write letters home, and to communicate with captors on matters of health and welfare.[6]

The explanation goes on to tell how the PWs must be careful when they fill out their capture cards and not to say too much when they write home. They are also told not to make any written or oral confessions, propaganda recordings for the enemy and, generally, not to do anything that will help the enemy's war effort.[7]

In fact, the whole manual (of 108 pages) helps fill in the sense of the code and, in addition, provides a basis for engaging in critical thinking about those times when one is a prisoner of war. It does more than that, to be sure. It also gives practical information about how to get along in prison (e.g., how to communicate via hand signals when the enemy forbids prisoners to talk to one another). Be that as it may, for our purposes, manuals like 21–78 show how teaching codes (intuitive thinking) and thinking about them (critical thinking) are not incompatible with one another. Putting it differently, it would be a mistake to teach the one without the other since without the codes there would be piecemeal confusion about what to do morally; while without critical thinking, the codes would be misunderstood, misapplied and, eventually, mistaken.

By now we have said enough about the codes. More needs to be said, however, about how critical thinking can be (and is) taught among military personnel. Another US Army manual is helpful in this regard, although this particular manual is only one of several that could be cited. The manual MQS III (January, 1983) is designed for a course titled 'Ethics and Professionalism' for advanced officer students.[8] The emphasis in the manual is to teach officers how to deal critically with problems rather than feeding them answers to these problems already examined critically by others. That is, the emphasis is method not solutions. In this spirit, officers are encouraged to learn how to reason by, for example, identifying options, ambiguity, vagueness and the like. They are also encouraged to get involved in the issues under discussion by using personal pronouns (e.g., by asking 'What would I do in this situation?'), reading about and discussing case studies and even engaging in some role-reversing exercises.

It is interesting that the case studies cited in MQS III correspond both to what we have called the fighting and the internal codes. It is also interesting that the specific topics range from whether it is proper to use prisoners of war to help clear a minefield (which the prisoners themselves have just laid down), across to internal security problems, all the way to ones concerned with sexism (e.g., where a female officer is assigned the task of getting the coffee for other officers).[9] It is as if the critical thinking sessions could not avoid ranging beyond military matters but had to include broadly social ones also. By discussing sexism, MQS III is not dealing just with what role women should play in the military, but with a comparable question having to do with the society at large. In this spirit, any discussion of racism is similarly not just about what goes on within the military establishment, but within the society as a whole.

Even ethical issues dealing with strictly military activities are not strictly military. When the military code of ethics tells us, in effect, that the preferences of the enemy need to be taken into account, this is so not just because wars have special rules mandating such an accounting. On the contrary, as we have indicated already, it is because such moral accounting is as much demanded in war as in peace. It is true that the parallel in and out of the military is not complete since the military is not run democratically as, hopefully, the society at large is. But the parallel is close enough so that the ethical rules of the military having to do with how it fights, how its members relate to one another, what its ideals are and even why anyone should serve in the military reflect what the society is about morally. Military ethics cannot be taught without getting involved in a discussion of larger issues sooner or later. Thus a good start to the touchy but seemingly necessary process of educating military personnel about 'civics' discussed in the previous chapter is to teach them military ethics.

Although military ethics is an excellent entry point for such discussions, the relationship between the study of ethics and civics is probably reciprocal. Not only does ethics lead to civics, but the more one learns about civics, the what, the how and the why of a society, the more likely the need for military ethics will be appreciated by military personnel. Also, the hope is, the more those in the service learn about these matters, the better they will serve their nation.

At this point the old worry reemerges that such wide-ranging discussion of social and moral issues will involve the military in politics and make it an instrument or affiliate of certain narrow political interests. But our discussion of how ethics should be taught shows how this worry can be allayed. Civic and political education need not be taught dogmatically. It need not be taught by an officer with a narrow view of things. It also need not even be taught negatively as it would if emphasis were placed, for example, on the evils of communism.[10] Instead, it should be taught much in the spirit of the moral critical thinking done in MQS III. By encouraging discussion, the expression of a wide variety of viewpoints, and the use of role-reversing techniques of teaching, the educative process is more likely to be effective than it would be if materials were presented dogmatically. It would not take long for those listening to 'a party line' to turn a deaf ear to it.

The civic and political educative processes would differ somewhat from those concerned with ethical critical thinking and the codes in that, in part, it would be more designed to inform. Military personnel would have to be told about the machinery of their government and the structure of their society and, if these tasks were done well, it would probably be because it was done by specialists in these matters.

Commanding officers could help here, as they could perhaps to a greater degree in presenting ethics materials, but they are not likely to be trained well enough to do this job on their own. Nor, given their other command responsibilities, are they likely to have the time to prepare themselves to make acceptable presentations.[11] Although it is probably desirable to have the commanding officers present during these presentations, it may at times be best not to have them around if free and open discussion is to be encouraged. However the details are worked out, the overall educational strategy, when it comes to teaching ethics, civics, and politics would look like this:

1   If nothing else can be done (because of an emergency) the various codes of ethics should be drilled into enlisted personnel. A military force that fails to take this minimal step will most likely end up in a moral morass. Even those military forces with strong officer traditions of honor and duty can no longer directly control lower-echelon personnel in this day when close-order formations and, therefore, close control over troops are not in fashion.

2   When more time is available, the codes should be supplemented with critical thinking. Such thinking serves not only to help explain why the codes are as they are, but shows the place codes have in the military. They also help fill whatever moral gaps are left open by the codes. Finally, such critical thinking helps begin the process of answering such questions as 'Why should we be in the military?' and, if it comes to that, 'Why should we fight?' Since the military codes and other rules not only guide the behavior of its personnel but reflect society's overall values, teaching the military's rules helps to show what the society stands for.

3   Beyond teaching ethics, answering the above questions requires both civic and political education. Unfortunately, many of those in the military services (in western nations) lack an adequate education in these areas.[12] And even if they do, they rarely, in their schooling, have discussed civic and political issues in relation to the military. So there is a need for such education and a need, we have argued, for undogmatic presentations of those issues that divide the political parties of a nation. These issues should be discussed in a manner that presents all sides fairly.

Having come out in favor of a liberal presentation of ethical, civic and political issues with the military, we are, nonetheless, not sanguine about the results of our proposals. The main reason for feeling as we do is that there is likely to be strong opposition to such a program of education from traditional elements within the military. The traditional officers will argue that these presentations will undermine authority and therefore undermine discipline. In turn, their lack of enthusiasm is likely to affect the final product negatively. It is also likely to be affected by a variety of other factors including the lack of good teachers, the shortage of time in an already busy military training schedule and the difficulties inherent in the process of teaching these materials to military personnel. Nevertheless, there is no reasonable option but to engage in such training programs. Both prudentially and morally, a military organization is better off doing so. Prudentially, for the liberal society at least, training its people in its accustomed non-authoritarian fashion is likely to create a more effective fighting machine. Morally the fighting machine that understands better what its purpose is, is less likely to commit atrocities upon others and its own people. So although the kinds of training we are recommending for those in the military may not achieve all that might be wished, failure to implement such training represents a worse option.

CHAPTER 4

The military and other institutions

I The military and the political institutions

We have already spoken of the tendency the military has to isolate itself from the rest of society. In some respects this isolation is salutary. Various forms of military activity are best done in relative isolation since, for one, these activities require the undivided attention of those who are involved in them. The seriousness of basic training would be compromised if it were conducted on an 8 to 5 basis, and if the trainees returned to their civilian homes each night and on weekends. For another, maneuvers and other military exercises are often dangerous to civilians. There is also an *esprit de corps* that can develop more readily when people work together, but yet apart from the rest of society. In addition, isolation is required often simply for geographic reasons. Major naval maneuvers, for example, can hardly take place within sight of shore.

In some other ways, however, especially when it operates as a state-within-a-state, the military can become too isolated. If military groups were not service organizations, there would be no problem. A monastery dedicated to purifying its members' souls might best operate as a small state-within-a-state. But an organization dedicated to serving others is another matter. It cannot isolate itself if it is to know clearly what the society expects from it. If, further, by the very nature of the task it performs, it is more powerful in some ways than the rest of society, its isolation can be a real danger.

Granting the possibility of danger, discovering the nature of the proper relationship between the military and the rest of society becomes a worthy endeavor. Clearly, that relationship cannot be characterized simply as the negation of isolation since some kinds of non-isolation are as undesirable as isolation itself. A military takeover can result in the society serving the military rather than the other way around. This is not to say that a military take-over is always undesirable. Sometimes corrupt and cruel governments cannot be overthrown without the active support of the military. Nevertheless, concerned as we are primarily with the relationships between military organizations as they operate within basically just societies, our problem is to uncover arrangements that permit the military to protect the society effectively and yet permit the society to protect itself from its protector.

In dealing with this problem, it is natural to turn to Samuel Huntington's *The Soldier and the State* for guidance. In this work Huntington charts, and by and large approvingly, the process which led to the professionalization of the military.[1] In the past and in many countries, he tells us, the military was run by officers who were essentially military amateurs. Usually this meant that since the rich and the powerful bought or were given commissions, the military, run by these officers, was in tune with the interests of the rich and the powerful. Such control of the military by one class of people is what Huntington calls subjective civilian control. Presumably he gives it this name since the control stems from the (subjective) interests of the class or group in power. 'Subjective civilian control achieves its end by civilizing the military, making them a mirror of the state.'[2] In contrast, he tells us, 'Objective civilian control achieves its end by militarizing the military, making them the tool of the state.'[3] He adds a bit later: 'The essence of objective civilian control is the recognition of autonomous military professionalism; the essence of subjective civilian control is the denial of an independent sphere.'[4] Still later he says:

> The one prime essential for any system of civilian control is the minimizing of military power. Objective civilian control achieves this reduction by professionalizing the military, by rendering them politically sterile and neutral. This produces the lowest possible level of military political power with respect to all civilian groups. At the same time it preserves that essential element of power which is necessary for the existence of a military profession.[5]

It is interesting here how closely Huntington connects 'professionalizing the military' with 'making the military sterile and neutral.' Such a connection may reflect history but it certainly is not logical in nature. Professionalizing the military is, as Huntington quite rightly tells us, a process of not filling the military with civilians who play soldier from time to time, but with personnel whose primary life occupation is military activity and whose standards of excellence also are primarily military. In this sense of professional, all of the major and many of the minor powers in the world have professional military organizations. But this is not to say that the military can maintain its professional status only if it completely avoids involvement in social and political issues. It may or may not be to the advantage of a highly professionalized military organization to become involved in politics at least to the extent of lobbying politicians for monies to run their organization. Whichever way it goes, at least professionalism *and* involvement in social and political issues are not contradictory concepts. Because they are not, it is possible to think of a military organization being both professional and heavily involved politically.

We have already argued in the previous chapter that there is nothing incompatible with having a military organization engage in politics internally by encouraging political discussion. After sharply separating discussions of social and political issues within the ranks (where a free exchange of ideas is encouraged) from task-assignments (where orders are carried out in a disciplined manner), we argued that the latter is not necessarily undermined by encouraging the former. In fact, we argued that in a liberal society, discipline may be maintained more effectively by giving military personnel organized opportunities to express their views on both military and social and political issues. We argued that these open discussions may be an effective way of getting military personnel to come to terms with why they should play the roles they do in the military.

But now, in this chapter, the question is whether the military should engage in social and political discussion and activity *externally*. In response, the extreme answers seem implausible. At the one extreme, where the military's involvement in political matters amounts to a take-over, there is an incompatibility between the military and the liberal society. The two, by definition, cannot coexist side by side. Yet it does not follow that the proper level of involvement by the military in political matters must be total abstinence. The military establishment deserves a fair hearing in the political arena as do other establishments such as health-care and education, since each provides services to the community that need to be explained and funded. Undoubtedly, these establishments differ. It is more difficult to conceive of the military, as against medicine and education, not being funded by the state. Also it makes little sense to speak of an educational or medical take-over of the body politic. Yet, if the issues that surround the operation of all of these establishments are complex, the traditional and preferred way of settling them is to foster general discussion about them. It would not do to discuss medical expenditures without involving medical personnel in the discussion. To argue that these medical people and their organizations should not be involved in the political process because they have vested interests in certain outcomes or because they are too powerful is beside the point. Of course they have vested interests. And in fact they may have become too powerful. But the proper response to such problems is not to emasculate this group's political influence, but to control it in some way.

The same point applies to the military. Potentially its power can be greater than the medical or the educational professions. That has been granted. Nonetheless, the need for open discussions of military matters is still present. The fact that the military can be a danger not just to the enemy but to the society that creates it does not diminish the need for open discussion. One would suppose just the opposite. The danger of a military take-over should encourage more discussion, not less. One would also suppose that part of the discussion that ensued, with the full participation of the military, would yield guidelines as to just how the military is to be controlled so that its power would not become greater than it should.

Certainly one such guideline would exclude the military, either on the institutional or the individual level from becoming involved in partisan politics.[6] This guideline would not prohibit the members of the military, both officers and non-officers, from voting as other citizens do. But it would prohibit the army, navy and the air force, and their personnel, from publicly taking sides with political parties, candidates or office holders. Since the military would probably have to perform its tasks during the ascendancy of all viable parties within a nation sooner or later, it would not do for it to become associated too closely with one side only. Such an association would neither be good for the military nor, more importantly, for the nation.

Non-partisanship is not a policy unfairly applied exclusively to the military. Those governmental administrators and personnel involved in community health are covered by the same guideline, and for the same reasons. The same is true of administrators and personnel involved in education. All of these people need to maintain some political distance since, like the military, they must eventually serve under all viable parties and, more basically, they must serve all the people of a society, and not just those who support them politically.

There is a problem here that although officially the military cannot support a political party, a party may officially support it. One party more than others may favor military spending and, in that way, come to be known as the favorite of some of the generals and admirals. Further, if as has happened in the United States, conservatives are attracted to the military as a career, and the conservative party supports the military, it should surprise no one if these two conservative elements developed a close relationship.[7] In a just society there is little that should be done to eliminate such a relationship since it is desirable to have the conservatives, as well as all others, express their political views. Further, it is undesirable to punish military officers because they happen to hold conservative (or liberal) viewpoints. However, what can be insisted upon is that officers maintain a neutral political stance to all the viable parties in their society insofar as they discharge their duties as officers. Minimally what this means is that they will not publicly endorse political candidates for office. Nor will they use any of the resources of the military to help one candidate or party. All that is clear.

What is less clear is what military personnel *can* do on the political scene for, after all, we are arguing that in some sense they and the institution they represent ought to be involved politically. Certainly their involvement should permit them to speak out on military matters and also to lobby for what they want. Within the government, the channels for so speaking and lobbying will vary with the type of government involved. In a country like the United States the military's political activities will permit it to speak to and lobby the legislature as well as the rest of the executive branch. A parliamentary government will offer narrower channels of expression. Whatever these channels are, the take-over and other dangers the military poses to a society are no excuse for suppressing the military's speech and lobbying activities since the military's views about military matters need to be widely heard and understood.[8] In fact, as important as some of these matters are to a society, the last thing we should want is to have them decided in secret by those who happen to be in power at the time. This is not to say that there are no reasons for maintaining military secrets. Rather, it is to argue for a policy of treating the military establishment as much like other establishments as possible. If, to return for a moment to the medical community, we permit it to exercise political influence, why should we not the military? It is a false model of how things should work in a just society to suppose that the military should maintain its professional purity by standing aloof from major political activity.

If our thesis is that the military is to be treated much like other institutions and establishments, how, it might be asked, is the society to be protected from the military? Has it not already been admitted that the military is more dangerous than the medical, educational and other high-spending establishments? Is it not more dangerous both by posing a take-over threat, and by gradually siphoning off monies that could be better used by the government and by the private sector of the society for providing human services?

Without denying that the military is more dangerous than other institutions, it should be clear that there is danger in whatever is done. There are dangers in isolating the military in an ethical cocoon all its own of honor, integrity, loyalty as well as in not isolating it. The dangers of isolation include, among many others, that the external threats to the society will not be appreciated since the military will not, in its isolation, have had an adequate opportunity to explain the nature of these threats. The dangers of not isolating the military are that it will become overly involved in politics. So there is no fail-safe procedure available short of eliminating the military establishment completely. Then there would be dangers of a different sort to be faced.

Granting these points, we are arguing that the kind of limited involvement in politics we are favoring represents the least dangerous option. No one would be naive enough to suppose that making the military 'go public' will significantly diminish the behind-the-scenes politicking that normally takes place. Nonetheless, under our limited-involvement model, at least some discussion of military matters is forced out into the open so that both the military and non-military come to understand one another a little better.

It might be argued that simply encouraging open discussion on broad military and spending policies is not very reassuring. Something more ought to be done to make certain that the monster among us does not eventually consume us. Quite right, more needs to be done both within and outside the government. As this chapter develops, some of the outside things that can be done will be mentioned. For now, however, our concern is to deal with the steps the government needs to take to minimize whatever dangers exist from the military. Aside from open discussion, one step quite properly taken by most western countries is to place civilians over the military as a whole and over each major branch of the military. Certainly another proper step is to develop bipartisan committees and commissions to review the military's actions and policies. These steps, in conjunction with those mentioned in the previous chapter aimed at integrating the military into the general society, can be pursued as vigorously as the need demands. Thus, if the military is seen to be more dangerous than the other institutions of the society, what would be needed is simply more of the same kind of treatment accorded to these other institutions.

It may seem otherwise to some since programs aimed at helping the poor are often seen as competing with those supporting the military and its high-spending tendencies. Favoring butter over guns, it is tempting for these people to silence the military politically and to see the military's involvement in this arena as sinister. It is also tempting for these people to disassociate themselves from the military on the ideological grounds that the military's disciplined group activities are inherently antithetical to individualism. It is as if those in the military should stand by ready to give their lives for the society they serve, should be grateful for what the society bestows on them but, under no circumstances, should they ask for anything. As some have put it, the military should be 'on tap, but not on top.'[9]

Ours is a different interpretation of how things should be. To be 'on tap' is a bit like the child being told to 'speak only when spoken to.' Such a policy of treating any individual or group is at least partly an isolative one. It forbids active advocacy, the option we are recommending. On the other side 'on top,' a policy sometimes called fusionism, insists that there are no distinctions between the political and military functions of government at the top.[10] It is in order to avoid having the military on top that some advocate that it should be merely on tap. But this is surely a false dichotomy. What is needed is not the political emasculation of the military as just being on tap, and not its glorification, but active participation hedged by checks and balances. The military should be viewed as a legitimate part of the society, an institution that has something to say to all the society's members, and as such an institution to which people should listen. Again, if it poses special dangers to the society, then special checks and balances ought to be arranged within the political process.

However, these checks and balances need not be restricted to those political processes taking place within the government. They can and should include other institutions such as the mass media.

II The mass media

The mass media are immensely more powerful than they were even a few years ago, thanks to television. News about military events, policies, purchases, appointments no longer has to be read about to be known about.[11] Of course the power of the media to portray events and to act as critics can be muted by censorship. What is more, that power can be turned around 180 degrees so as to support the military rather than be an honest critic of it. But in a society where the media has had a tradition of being free, there is little doubt that it can supplement the watchdog functions of those in the political sphere.

In this sense, the media's responsibilities are one sided. Their task generally is to look critically at all of the institutions of a society.[12] Often they are criticized for doing this in an overenthusiastic manner. Additionally, they are admonished for not presenting a more balanced account of what goes on within a society, that is, for not telling the good news along with the bad. But, on the view we are defending, it is not the sole or even the main function of the media to present a balanced account. The media can certainly be irresponsible if they distort, suppress and fabricate their reports. They are also irresponsible if they fail, at least from time to time, to remind us that a particular institution within our society is serving the society well. But somebody doing something right is what we normally expect. In spite of widespread cynicism about the functioning of the military, most things probably get done in most military establishments within acceptable standards roughly comparable to those reached in other governmental agencies and in major industries. If this is so, we should not expect the media to tell us constantly about those things that we expect to happen. This is telling us about something we already know.

Forming as they do a powerful institution independent of the government, the media have a special and heavy responsibility to act as critics. The military is, after all, part of the government, and many governmental agencies and officials, although they are not a part of the military, work closely with the military in ways that can cloud their sense of objectivity. Certainly local media can have their judgments similarly clouded. A newspaper might favor a locally produced weapons system since it will be good for local business and, indirectly, good for it. But, as a whole, the media are not so tied to such pressures as even opposition legislators may be. Being relatively independent of the government, the military and certain local pressure groups, they are in a position to distance themselves from the military and so act as critics. They can prevent the military from hurting itself and the rest of society by playing this critic's role.

Instead of being the enemy, then, the media are the loyal opposition. At least that is the way it should work in theory. In practice the media may fail in one of two ways. First, they may fail by simply not doing their work for any one of several reasons. They may simply not care, feeling that the military is less newsworthy than are medicine, crime and Hollywood. The media may fail because they may lack the expertise, even if they do care.[13] Or they may be afraid to look into such matters. Being critical of the military may not seem patriotic to some advertisers. Or the government itself may put pressure on the media to avoid discussing military issues. For whatever reason, such failures will make the military comfortable in the short run since it will not be receiving criticism but, in the long run, it will likely make it less able to perform its proper functions.

Second, the media may fail by criticizing the military unfairly. Here we need to remind ourselves that the media are not composed of homogeneous entities. At least in most western countries, the media are composed of various independent units communicating through television, radio, the newspapers, magazines and other vehicles; and managed and operated by people with widely differing political and social viewpoints. Some of these viewpoints might be anti-militaristic in principle as in the case of a pacifist publication. Some might be anti-militaristic in the narrower sense of opposing only the military of the society in which they work because they wish to overthrow the government and the social order. From our perspective, there is nothing wrong with such criticism. These people have the right to criticize the military since they are probably doing the society, and possibly even the military, a favor in expressing their criticism. However, if their criticisms are based upon factually incorrect information or are overly strident, those who are the targets of this criticism have a right to reply. The same point can be made in utilitarian terms. If there is utility in having the media criticize the military, there is utility in having the military reply because the back-and-forth discussion should yield information and give perspective to all those involved in the decision-making process and, thereby, help them to make sound decisions.[14]

It might be thought, in reply, that even if it is important for the society to have information about the military widely disseminated, there is no need for the military to become involved in this process. After all, if certain segments of the media are against the military, other segments are for it. So if the anti-military media misrepresent the facts, the pro-military segments should set things straight. Not only is there no need for the military to get involved in a media quarrel, but such involvement simply distracts the military from its appointed task of military preparation. True professionalism, as Huntington might have put it, requires dedication to military practices. Fighting battles with the media is a move away from such dedication just as is playing politics.

Before replying to this objection it is well to remember that verbal attacks directed against the military are not restricted to those coming from the media. Putting it in terms of rights theory, the right to freedom of speech is not restricted to the media in most western nations. All citizens have that right.[15] In more utilitarian terms, people, all of them, not just media people, have been given that right in part to guard against the consequences that the media themselves might become corrupted. Such corruption might take the form, for example, of having the media controlled by certain classes of people so that they are used to express mainly their own views and not necessarily the views of others in the society.

Our reply to those who object to giving the military the right to speak up comes down to this. Since the media, other institutions and any number of groups can unfairly criticize the military, and since it is not clear that even the friends the military has among the media can redress the unfairness, it seems only fair to have the military try. Although the parallel is not exact, it is as if the military, at least insofar as military matters are concerned, needs to play the role of keeping the media and others honest about how they do their work, just as the media have the role of keeping the military honest about how they do theirs.

One of the consequences of this quasi-adversarial relationship between the military and the media for which we are arguing is that certain kinds of censorship will be forbidden. It will not do for the military to argue as they did in the Falklands and the Grenada incidents that media personnel should not be allowed on the field of battle because of the danger involved to them. Danger is certainly present in events like this but, presumably, the media people are well aware of that and are willing to live with that danger. They have a dangerous and important job to do just as do military personnel in battle. For the military to show such great concern for the safety of media personnel to the point of excluding them from the battle scene seems, therefore, a bit disingenuous.

This is not to say that there are no reasons for excluding the media from certain battle situations. There are operations, morally legitimate ones, where having anyone present other than those carrying them out constitutes interference. When the media get in the way of the work that the military must do, there is at least a presumptive reason for delaying their arrival.[16] Commando operations come to mind here as do the initial phases of somewhat larger operations. Still, in these latter cases, selected media people ought to be allowed on the scene as soon as possible.[17] More than that, it is hard to believe that any really major operation could not find a niche for at least some media people where they would not get in the way.

Excluding media people and perhaps censoring their speech is also justified when it comes to highly secret information. Again the criterion of exclusion is interference, now interpreted more broadly. To publish information about a secret weapon, or about the timetable concerned with troop movements, is to give intelligence to the other side and, in that way, to interfere indirectly with the military. Such interference will lead the military to change plans, possibly accept higher casualties and even suffer defeat in battle. It cannot be countenanced.

The reasons cited thus far for limiting the work of the media suggest that there will be fewer limitations on them during peacetime. In peacetime, we not only expect the media to continue their coverage of the military, but to do so without carrying a heavy burden of information being withheld from them. There is also the issue of who decides what needs to be withheld. Before dealing with that, we need to come to understand the power of the media a little better.

As we have been pointing out, the military can control the media by excluding them from their wars and war games, and by keeping lots of secrets. However, once the media work their way through whatever censorship screens have been thrown at them, they have a relatively high degree of freedom in doing what they want with the information they have uncovered. Occasionally, we need to be reminded of the obvious, that the media control the media. This means both that they can report and comment on much of what they have uncovered even though the military may not want them to, and that they need not report or comment on what the military wants them to. Although the military can withhold information, the media, in turn, can withhold access. This power of withholding can take the form of simply ignoring what the military says, putting what they say on the back pages of a newspaper or placing it in a bad light. These options mean that the military can at most manipulate the media. It cannot control it easily. Often when people talk of the military manipulating the media they imply that the military, in western countries at least, is in a very powerful position with respect to the media and the society at large. Quite the opposite is the case. To a certain extent the military can orchestrate certain information it releases so that it has either maximal or minimal impact. Unpleasant information can be released late Friday night so that it can conveniently be forgotten over the weekend. But since the instruments of the media are not in their hands, and in many cases are in the hands of media personnel who are not in sympathy with the military, the sense of manipulation referred to above certainly cannot be anything like the control of someone who manipulates his puppets.

So, in its own way, each side can undermine the other and, in the process, turn a quasi-adversarial relationship into one that is simply adversarial. Roughly speaking, the rules of the former relationship are that each side, the military and the media, will let the other do its job with a minimum amount of interference.[18] More than that, each side should cooperate to help the other do its job better. On the military side, that means that even if it is an unpleasant experience to let the press look at its dirty linen, it should see that such openness will help it to maintain credibility with the community at large, and force it to correct whatever its errors are. On the media side, it should see that giving the military access to the media is part of its responsibility to society. It should not suppress information and commentary about an institution that is especially hard-pressed to present its case to the society it is trying to serve.

It is one thing to say that the military should maintain more openness than it has traditionally practiced, it is another to make suggestions that might encourage this openness. The problem is that if the military alone decides what is a military secret and which operations the media can attend, then it becomes impossible to sort a legitimate case of withholding information from a cover-up which protects either the military as a whole or certain people's military careers. Making the decision process less one-sided would help some. Having and maintaining a committee made up of media representatives (chosen by the media of course) and representatives from the military and government to establish guidelines for secrecy would at least put on record how things ought to be done. If on some occasion the military violated the guidelines, the media could voice their objections through the committee as loudly as they might wish. Under such an arrangement, many events and much information might still be wrongly labelled secret by the military and their civilian leaders, but at least some of the media's complaining would have more bite to it if it were based upon a violation of standards agreed to by the military itself.

Obviously the media can do more than complain loudly to those who view, hear and read them. If they have been mistreated by the military, they can complain to the government and in the process exert strong political pressure to encourage the military to permit the media to perform their proper tasks. They can also be more aggressive in searching for certain kinds of information, although here we run across delicate moral and legal problems concerned with degrees of aggressiveness. There certainly is nothing wrong with publishing information not widely disseminated but available in military publications. Presumably that information is known to potential enemies so there can be no good reason for not making it available to the society as a whole. The delicate issues arise when the media releases censored information that has been obtained from some inside source. Assuming for the sake of the argument that the released information is not properly secret but has been classified that way in order to protect incompetence or bungling, the release would seem justified. It is true that encouraging this practice would encourage subversion of a certain type. The malcontents in various military positions would be encouraged to turn to the media whenever they have a disagreement with their superiors. Every inferior would, according to our recommendation, be a potential media spy and, as a result, no one in the military would trust anyone. Without denying that the loss of at least some trust is a loss within the military community, not giving the media access to information through irregular sources is a greater loss for the community at large. It must not be forgotten that even if the media have a right to complain after the fact when they discover that the military has been withholding information without good reason, and even if the media and the military have come to some agreement about what information should be withheld, the power to withhold still rests with the military and the government. Being very strong, that power will inevitably be abused in ways that might not ever come to light if some inside source were not to reveal to the media what is happening. To put our point in the strongest possible terms, if the media uncover abuses, they have a duty to bring these abuses to the surface. A corollary of this point is that those media spies, so called, also then have a duty to help the media in this regard.

It should not be forgotten that we are concerned here only with abuses of power by military personnel and their governmental allies. If the information leaked to a certain media outlet is a genuine military secret then that is another matter. Now the presumption is that the leak interferes with the proper function of the military whose function is to defend the nation, and the rights of the media to say what they want do not apply here. If, to take another kind of case, a media outlet receives genuine military secrets 'out of the blue', then that outlet's duty is to return those secrets to their source without making them public. The media outlet may, as a result, have to forgo an exclusive story and thereby diminish its profits. Whatever the media's rights may be, they certainly are not so unlimited as to allow profits to override the security and welfare of the nation as a whole. Undoubtedly there will be borderline cases as to whether what is revealed is a military secret. Presumably when these cases arise, decisions should be made one way or the other by some independent court or judicial authority. But the fact of the existence of borderline cases cannot be used as an excuse to deny the media the opportunity to play its proper role as a critic of the military.

There are, of course, other independent institutions that also play the critic's role. Perhaps the foremost among them are the church and the university. In one way, the former can act as the media do. Although various churches do not have the investigative power of the media with their armies of reporters, the churches, like the media, can reach many people. It is quite the opposite with the university where the ability to reach large numbers of people is lacking, but the research power is great. In addition, there are academic institutes that are not necessarily part of some university that also engage in research on military matters.[19] All of these together with informal peace and defense groups (e.g., veterans groups), quite properly play their role in watching over military activities and rendering judgments over what the military is doing. Granting the inherent dangers of having a military organization within a society's midst, it is tempting to say about these groups, the more the merrier. Thus if one group, for example, the media, fails to be alert enough to observe a dangerous trend within the military, some other group we would hope would pick up the slack.

III Military industrial complex

In discussing the relationship between the military and the rest of society in this chapter, we have dealt so far only with those institutions that can and should pay quasi-adversarial roles vis-à-vis the military. We now turn to business and industry where historically the relationship between certain segments of these institutions and the military has been more cooperative than adversarial. The cooperation has been so cozy, in fact, that the expression military industrial complex has come into fashion to characterize it.

In Chapter 1 we argued that preparations to fight in self-defense are morally justifiable. Defense, however, requires armies, and armies in turn need to be supplied with large amounts of specialized and sophisticated equipment. Industries devoted to supply these needs must, therefore, be built. Further, the production of such equipment requires teams of highly skilled and specialized technicians, scientists and engineers who cannot be brought together quickly and easily. The level of technology is such that even production line workers must be more highly skilled than average. Finally, such military production requires huge capital expenditure for plant and machinery.

Given these facts, it is not surprising that few businesses find themselves in position to produce these military supplies.[20] Also, since most of this work involves secrecy, and research and development requires close cooperation between members of the military establishment and corporate suppliers, it is almost inevitable that close and enduring ties have developed between them. This condition is exacerbated by the fact that the mid and late twentieth century has been in a near-permanent state of tension and unrest. Military forces have been maintained at permanently high levels of readiness for over 40 years, and the pressures of international rivalry fuel the search for ever more potent and sophisticated weapons.

Any number of unhappy results have been attributed to the existence of this military industrial complex but, for our purposes, two stand out as especially troublesome.[21] The military and the defense industry share an interest in pressing for ever-increasing supplies of ever-more-powerful and expensive weapons. Their interests are especially potent because the symbiotic relationship between the military and its corporate suppliers gives each greater influence than either would enjoy alone. But this shared interest in a high level of weapons production may often be at variance with the best interests of the nation as a whole. The complex may, for example, take steps to keep international tensions high or at least to create that impression. The complex knows well that the best way to insure success in budgetary deliberations is to be able to point to some brewing crisis which requires a strong military presence or response.

The second difficulty is arms sales abroad.[22] Since the domestic market is limited, corporations will naturally turn abroad in order to increase sales and diversify their markets.[23] Both government and military officials share an interest in fostering such sales. Increased production lowers unit costs. Arms sales can also cement alliances and insure dependency of client states because of the continued need for parts and updated equipment. These exchanges also help with balance of trade problems and preserve jobs. The problem, of course, is that such sales serve both to increase tensions and make ensuing wars more destructive. Further, precious foreign exchange spent on arms is then unavailable for other purposes such as caring for the poor.

We do not see how, given the permanent state of world tensions, the development of permanent arms industries with close ties to the military can be avoided. To those who claim that the alliance of military and industry is a byproduct of the capitalistic drive for profits, we note that similar arrangements exist in nations not known for their strong devotion to capitalism. We note, further, that there are structural features of such systems that will tend to cause similar problems wherever munitions industries exist. Arms suppliers face the same constraints as any other industry, with a few extra problems thrown in. They require a regular return on capital. They need a constant schedule of production to keep plants and workers occupied. They need work for their high-powered development staffs.

These problems are shared with other industries. The added difficulties they face are that they have essentially one customer, the government, that can increase orders, cut them back to nothing, or demand entirely new and untested products on a whim. It is, in addition, very difficult for these industries to diversify. The technology and equipment required for military use are not easily transferred to civilian requirements.

It is our position that the difficulties posed by the existence of the military industrial complex cannot be eliminated. They can only be mitigated. No doubt, it is tempting to argue that the problems could be avoided by simply dismantling the arms industry. When crises develop, it might be argued, nations could simply convert civilian industries to military production as they have done in the past. Sadly, this option is hopelessly unrealistic. The advanced state of modern military technology and the requirement for putting it into production are such that plans cannot simply be shelved and dusted off when needed. Maintaining the technology required for effective military forces requires keeping plants in operation and staffs in place. Modern wars are not likely to cooperate by beginning at a leisurely pace and spreading gradually, giving time for conversion. Finally, the permanent state of world tension requires the active maintenance of military forces at a high level of preparedness.

Nationalization is not a solution either.[24] The socialist economy of Russia has produced a military industrial complex with many of the features found in the US version. For example, equipment costs have risen there just as they have in the US.[25] More fundamentally, however, nationalization does not eliminate the structural features that require regular employment of staff and constant plant operation, or that necessitate close cooperation with the military. Even socialist managers realize that they enjoy a close community of interest with the military.

While there are no total solutions, we believe that a number of steps can be taken to mitigate these problems. For one thing, there are oversight groups, such as the General Accounting Office in the US government, that can be and are charged with examining expenditures and effectiveness of the arms bought. These groups are often well-informed and well-prepared. Their effectiveness is illustrated by the fact that critics of governmental spending in the US gain much of their information from them. These groups can both be allotted greater independence and given a more prominent role in making decisions, should more need to be done to control the military.

A second step is to come to realize that difficulties with the military industrial complex are symptoms rather than direct causes. Arms purchases are the result of foreign policy decisions and perceptions of international crisis. It is true enough that members of the complex may attempt to manipulate these decisions and perceptions to serve their own interests, but they cannot be said to have created them. They did not invent the rivalry between the US and the USSR, for example. The best way to keep their influence from getting out of hand is to subject their claims about world crisis to clear-headed analysis via the institutions discussed already in this chapter both in and out of government.

A third step is to recognize frankly the needs of these industrial groups and cooperate openly with them to meet their reasonable requirements. The difficulty in the past has been that they have been left to get by as best they could and did so by whatever means came at hand. A frank and open policy would help to minimize subterranean abuses here and thus regularize production schedules. We should acknowledge the special circumstances of these industries that have essentially one market and cannot easily diversify into other areas.

Although these measures might help significantly to mitigate domestic problems, they very likely would be only of marginal help in controlling the international market in arms sales. Too many nations are involved here, and they have too much at stake for unilateral action to have much of an impact. Once more, it needs to be recognized that international arms sales are symptoms of problems rather then the problems themselves. While such sales may certainly make the problems worse, they do not create all the rivalries that exist in the world. Nonetheless, we believe that even here some measures can be taken to mitigate the problem.

A first, necessary, step is to recognize that international co-operation is required. Many nations have arms industries and nearly all depend on foreign trade to keep them afloat. No solution can be expected to work that does not take these very important economic stakes into account. For this reason, measures should be taken to attempt to control the arms trade rather than bring it to a halt altogether. Nations could agree that certain types of weapons would not be produced for export – planes with long-range bombing capacity, for example. They could also agree to put ceilings on performance and destructive capacity of weapons produced for export. These measures have a reasonable hope of acceptance by interested parties and could serve to put a cap on arms sales while not eliminating them altogether. In addition, international agreement could be effective in meeting the all-too-important problems of financing. Agreeing not to grant certain forms of credit or refraining from granting credit to nations in economic difficulty would go at least some way toward resolving this problem.

A second useful step is to encourage joint production of weapons systems. This makes military sense because allies benefit from using similar equipment. It saves on production costs as well and could be formulated to give each partner a slice of the pie.

In this area as much as in any other the best is the enemy of the good. The search for complete and final solutions that ignore the pressing reality is likely to have no effect at all. The only way to begin to deal adequately with the problems posed by the military industrial complex is to recognize that they will not disappear so long as the world is as it is. By understanding the conditions which cause the problems to arise, however, we can take at least some small steps to control them and minimize the damage that they may cause.

PART TWO:
ISSUES IMMEDIATELY
PRECEDING A WAR

CHAPTER 5
Just cause of war

I Just war theory

There has been a great deal of debate in western thought about what conditions need to be satisfied before going to war can be claimed to be morally justified. The various aspects of this debate form part of what is called *just war theory*. Telford Taylor has distilled the main conditions that need to be satisfied here and also indicated where the main areas of controversy lie.[1]

Taylor notes that the first condition is that the war must be declared by a legitimate authority. In the present scheme of things, only sovereign states are legally entitled to issue declarations of war. The next condition is that the war must be initiated for a just cause. This is normally interpreted to mean that the war is only justified if it is a response to some prior misdeed committed by the adversary nation. In addition, to be just, wars must be waged with the right intentions. The notion of 'right intention' is somewhat mysterious. Apparently, in Taylor's understanding at least, such intentions are the goals or ends which belligerent nations seek to attain in fighting a war. The relation between just cause and right intention is complex and difficult to understand. A just cause of war might be an act of aggression caused by another nation. One's intentions may then just involve negating the ill-effects of that aggression or creating conditions that will prevent aggression from recurring. But, as we argue in Chapter 11, any number of goals may be formulated and sought once war has begun. Some of these goals will derive from the initial just cause of the war while others, such as seeking to correct injustices of the distant past, may be unrelated to the present just cause. Finally, the war must be waged using the correct means. Rules purporting to govern the correct conduct of war abound. Most military groups, from the time of the Greeks at least, have developed standards of moral conduct that are often quite strict and demanding. There is no universal agreement on these standards beyond the consensus that some should be followed. Nevertheless, the point of including right means along with the other preconditions of a just war is that a nation would not be entitled to engage in a war that it intended to fight using immoral means. This would be so even though it had just cause for going to war and had noble intentions. Some would argue, for example, that a nation could never justly participate in a war fought with nuclear weapons or in one that necessarily involved killing large numbers of innocent people.

Apparently, though Taylor does not explicitly say so, the claim is that no conditions other than the above four are necessary and that no one of these conditions, by itself, is sufficient morally to justify going to war – including that of just cause. Therefore, even if a nation has just cause for going to war, it would be morally wrongful for it to do so if it has such wrong intentions as gaining extra territory. Responses to encroachments on a nation's territory or overriding its sovereignty, by themselves, are not adequate to morally justify going to war. The other three conditions of a morally justified war must be satisfied as well. We will examine issues of the just means of fighting war in detail in Part III (Chapters 7–10) of this work. Issues of right intention and their relation to just cause are discussed in Chapter 11. The just cause condition, however, is prior to the others in that they are partially derived from it. The goals sought by war, for example, must include rectification of the wrongs which caused it. The just cause must also be temporally prior. If the just cause does not first exist, no authority may legitimately declare war. Also, no set of goals, however noble or desirable, may be sought by war unless there is a just cause. That is to say, the claim is that it would be wrong to start a war, even to establish heaven on earth, if there were not some wrongful act whose evil is of sufficient magnitude to institute war. Because of this priority, we will devote the remainder of this chapter to analyzing the problems of just cause, the area of the topic which has been the subject of the most intense controversy. We will, however, also have some things to say about the relation of just cause to the other three conditions of just war.

II The just cause

The central idea behind the idea of just cause is that wars are responses to prior wrongs.[2] The just war must be reactive or negative in nature – not positive. The implicit idea is that crusades, efforts to make the world better, or the pursuit of ideal goals, are not sufficient as just causes of wars. A nation would thus not be warranted in starting a war to make a good world even better or to spread its own version of a utopian society. This premise is so deeply embedded in just war theory that it is difficult to find an explicit discussion of it. But, it might be asked, why should wars be only reactive in nature? Why cannot wars be useful instruments for creating a better, happier, or more just world?

This presumption is a sound one and can be supported easily. The key lies in the main feature of war itself. It necessarily involves the destruction of human life and well-being, often on a large scale. So it is essentially an instrument which causes great harm. Use of this instrument can only be justified if employing it achieves something of such great value as to outweigh its harm. It is difficult to imagine anything more important than the prevention of the destruction of human life and well-being. The harm of war can only be justified if it prevents an even greater harm of the same sort, because it is rare that any positive goals will exceed it in value. The cure of disease, the establishment of world peace and harmony, or the achievement of universal human happiness come to mind as possible candidates; but it is improbable that the crude instrument of war is capable of achieving these goods.

Our point is not that preventing harm is more important than achieving some good. Rather it is that the harm of war is immediate and quite certain, while the goals that might conceivably justify its use, other than forestalling death and destruction, are vague and elusive. Even if one's war is successful, these goals are still unlikely to be attained. It is hard even to know what steps might be required to achieve them.

From another perspective, if present efforts to improve the world fail, we can always make another attempt at a later date. But lives lost or ruined next week cannot be repaired later. If today's chance to prevent such harms is missed, it is lost forever. Thus war as a desperate measure seems suitable only for desperate circumstances such as those concerned with life and death. Also, even if there were reason to believe that utopia could be achieved by war, it is likely that it could also be achieved by some other means. When human life and well-being are directly threatened, though, usually the only way of meeting the danger is by military means. It is highly unlikely that Hitler, Idi Amin or Pol Pot could have been successfully deterred from their schemes by any means short of war.

Lastly, whatever changes in people's lives war brings about will be due to physical coercion. War is the ultimate instruction of coercion. Goals that are otherwise clearly worthwhile may not be so where the means used to achieve them involve physical coercion. However compelling one's vision of a utopian world order may be, it is unlikely that actively seeking it will justify use of the methods of physical coercion.

III Aggression

We are in agreement, then, with the conventional view that wars should not be used as positive instruments to serve great causes. Instead, they are normally justified as reactions to great wrongdoing. In this century the conventional wisdom has been that the one wrong of sufficient magnitude to justify war is aggression. The thinking has been that only aggression is sufficient to justify going to war and, further, that the crime of aggression is *always* sufficient to justify war. At this point we part company with conventional wisdom, as others have also begun to do.[3] Both of the above assumptions are at best oversimplifications. To support our doubts about these assumptions, it is important to try to understand just what aggression is and what makes it wrong.

One difficulty with understanding 'aggression' is whether it is purely a descriptive term, like 'killing' or whether it contains some normative content, like 'murder' so that aggressors can be said to be acting unjustly by definition. Michael Walzer, who has discussed the topic as carefully and sensibly as anyone, defines aggression as any *violation* of the territorial integrity or political sovereignty of a nation-state.[4] For him, then, aggression is a normative concept. A violation of territorial integrity would be a refusal to recognize the physical boundaries of the state, while violation of sovereignty would be a refusal to recognize the right of the government of the state to be the ultimate legal authority on all matters within it. These violations are wrong for Walzer; indeed they are the greatest crimes that can be committed against states. They are so because aggression strikes at the right of self-rule of a people and forces them to risk their lives in defense of their political culture.

Walzer is in agreement with the conventional view that an act of aggression *always* is just cause for going to war. He also accepts the usual convention that the crime of aggression is a crime against states. The right to be free from aggression is a right held by states rather than by individual persons. States possess this right, in his view, because they develop from the common life of a people and because all states must be presumed by outsiders to protect that common life. Of course, individuals will be harmed by aggression insofar as it undermines their right of self-rule and forces them to risk their lives in making a response. Even though they are harmed in this way, they are not holders of the right.

Though these views are in full agreement with international law and common belief, we shall argue later on that they lead to some difficulty. A more useful ploy, we will argue shortly, is to view the wrongfulness of aggression as the direct result of the harm that it causes to individual lives and well-being.

One difficulty with Walzer's discussion is that it is not clear why aggression, understood in his sense, is always wrong. When allied forces invaded Nazi Germany during World War II, they clearly violated its territorial integrity and ultimately its political sovereignty as well. Yet, common sense is reluctant to think of such actions as wrong. In fact, Walzer himself endorses the invasion.[5] Worse, we bridle at calling it aggression at all. Perhaps our reluctance is caused by the general feeling that the invasion was justified and that 'aggression' should only be applied to wrongful acts.

However, the most fundamental problem for Walzer is that territorial integrity and political sovereignty can be threatened by non-physical, non-military means. Political sovereignty can be threatened by resolutions adopted by the UN or by violations of treaty agreements. Also, territorial integrity can be violated in any number of ways short of warfare or the use of violence as when a nation violates another nation's fishing rights or sponsors illegal settlements within its borders. These cases are in conflict with our ordinary view that aggression must be a military activity or at least involve physical coercion.

Furthermore, and most importantly, it is difficult to see how any of these actions will necessarily provide full justification for going to war. Violation of fishing rights, for example, might be such a justification but only if it threatened the well-being of an entire nation, as may possibly have been the situation of Iceland in its recent

'Cod War' with Great Britain.[6] Lacking such grave consequences for individual persons on a major scale, it is hard to see how violations of territorial integrity will always justify the full rigors of war.

Walzer can also be taken to task, and he has by such writers as Charles Beitz, Gerald Doppelt and David Luban, for presuming both that all states have the right to be in power and that citizens are somehow obligated to respond to all acts of aggression. The problem with the first belief is that it is unlikely that states grossly violating the human rights of their own subjects have any moral claim to retain power. An attack directed against such states then would not be morally wrong. Furthermore, they would not have the moral authority even to resist such attacks.[7] This opens the second avenue of criticism which is that unjust states will not have the moral authority to compel their citizens to go to war to resist acts of aggression. They, of course, will have substantial resources to coerce them into fighting but that is not the same as having the moral authority.[8]

These critics often go on to claim that just states are legitimate and thus have the right to resist all acts of aggression.[9] We believe that these views are wrong. For one thing, it is difficult to imagine why even a state that practiced the most lofty ideals of justice would have the right to go to war to resist any or all acts of aggression. As we noted above, it is hard to see how a state would automatically be justified in going to war over violations of fishing rights – an aspect of territorial integrity. Furthermore, justice is always a matter of degree, of more or less. Would a just state have no right to resist aggressive acts in service of establishing an even more just state? Any attempts to ground the wrongfulness of aggression in the rights of states, even just ones, will likely fall prey to similar problems. Given these difficulties, it is tempting to return to the original common-sense view of aggression as initiating war. The normative load of this understanding is carried by the assumption that it is always wrong to start a war. Yet, this view is not entirely satisfactory either. Among its problems are those related to the issue of what it is to begin a war. 'Firing the first shot' is not particularly helpful, both because it is often difficult to discover just who fired the first shot and also because the exchange of shots need not lead to all-out war. China and Vietnam have been engaging in artillery duels for some years, yet neither seems anxious to engage in all-out war.[10]

Another view is that the first nation to send troops across international borders in significant numbers is the initiator of war. It is fairly easy to discover who dispatches troops first, and it is rarely the case that troops can violate international boundaries without war following. The difficulty is that the actual flash-point of fighting may occur only after a long period of increasing tension and mutual provocation. It may happen that one nation is responsible for initiating a build-up of tensions, while the other is finally the first to engage troops. In these cases it is difficult to determine who is the actual cause of the ensuing fighting. Deciding who was the first to commit troops, therefore, is relatively unhelpful in attempting to settle this issue. The nation that sends troops in first may actually have been the victim of hostile gestures and intentions of the other. It may be that both parties are equally to blame or even that both are victims of tensions and maneuvers generated by yet other nations. If we wish to retain the idea that aggression is always morally wrongful, we must reject the attempt to define it as 'initiating war'. Sometimes the act of initiating war, as the above examples show, can be morally justifiable or at least neutral from a moral standpoint.

It is apparent that there is a substantial amount of confusion about the term 'aggression' and no clear consensus on how it should be used. A useful solution is to reserve the term 'aggression' for the wrongful initiation of war. Rather than look for a single definitive cause of war, such as 'firing the first shot' or being the first to commit troops, it is also more useful to view the initiator of war as the party that is the preponderant causal agent. We will claim, finally, that wars are initiated wrongfully when they are *not* undertaken in response to significant threats to the lives and well-being of individual citizens. They are wrong because the death and destruction necessarily caused by any war can only be justified by preventing even more death and destruction.

IV *Other just causes*

Recent discussions also demonstrate that there are just causes of war other than response to aggression, and that sometimes even initiating war is morally justified. There are instances where a nation that does not strike first, or allows enemy war preparations to continue unimpeded, runs a serious risk of being unable to respond adequately when war does come. It seems perverse to insist that even in these circumstances the hard-pressed victim must await the first blow on grounds that it is always wrong to initiate war. Even well-intentioned nations may find this more of a burden than they are able to bear. In addition, there is little justification for asking them to sacrifice their own citizens and increase the chance of military defeat for the sake of an abstract principle. It is quite true that such justifications for warfare will be misused by rogue nations. But a perfectly good argument should not be dismissed simply because it may be misused. In any case, it is unlikely that rogue nations will be seriously deterred if such arguments are unavailable. They will continue with their plans with or without the fig leaf of appeal to principle.

Nations clearly are justified in resorting to the first strike where

1   it is extremely likely that war is imminent anyway;
2   they will suffer great risks of destruction if they wait the first blow; and
3   there are no other means available to avert the crisis;
4   their antagonist has no just cause for war.

We recognize, of course, that firm evidence and reasoned deliberation for accurately making such judgments are usually difficult to acquire. It is extremely difficult, for example, to determine whether war is inevitable or how much suffering accepting the first blow may cause. As a result, it is tempting in such cases to say that it is better to err on the side of caution. The difficulty is that when the above four conditions apply, it is *not* better to err on the side of caution – the loss suffered from doing so is simply too great. Of course, in many instances of threatened war it will be apparent that the stakes are not this high. There are some wars, such as in the Falkland Islands, that both nations could well have afforded to lose. There are other instances, as the attack on Pearl Harbor in 1941, where accepting the first blow was actually beneficial in galvanizing a nation into action. The way in which national leaders get into difficulty, and cause great hardship both for themselves and for others, is by treating all instances of threatened war as approximating the conditions which justify a first strike. As we pointed out in the discussion in Chapter 1, it is all too easy for nations to become neurotic about their security needs and to stretch the web of security sensitivity beyond all reasonable bounds. Long-term threats or threats to the periphery of one's security web are not sufficient to justify the cost in human life and, in any case, are all too often simply the pretext for other less laudable aims.

We have been arguing throughout this work that no nation can remain completely indifferent to the welfare of the citizens of other nations. Their lives and well-being must matter to all. Particularly when in grave danger, they will matter a lot. One implication of this point is that sometimes nations will have the obligation to initiate war on behalf of these citizens from other nations. A strong case can be made, for example, that other nations had an obligation to come to the aid of the Cambodian people when they were at the mercy of the murderous Pol Pot regime in the 1970s. In a similar vein, Milton Obote of Tanzania had excellent reasons for disregarding the principle of non-intervention when he rescued the people of Uganda from the rule of Idi Amin.[11] Obote clearly violated the territorial integrity of Uganda and just as clearly initiated the war with Amin's government, but there are few who would argue that his actions were not justified. So going to war on behalf of large numbers of people in other nations is sometimes justified and may even be a moral obligation.

Obviously, intervening in the internal affairs of another sovereign nation is not something that can be undertaken casually. This is not merely because nations commonly interfere in the affairs of others mainly from selfish or imperialistic motives. Governments of nation-states have the primary responsibility of looking after the well-being of citizens within their borders. Furthermore, they have the task of serving as representatives of these people to the world at large. Intervention into another nation's internal affairs is thus likely to be met with great resistance and be costly in terms of human and material destruction. More fundamentally, such intervention under-cuts the institutional arrangements that have been made to look after individual human beings. In addition, it undermines the freedom of peoples to devise their own modes of government and live as they wish. Thus such intervention always carries the danger of violating the right to self-determination of national groups and of destroying the governmental structures which provide order and security for them. It can be justified only when special circumstances exist. These circumstances should include the elements that large numbers of people are being deprived of their lives or well-being, these deprivations are permanent and ongoing, the people clearly are unable to help themselves, and they clearly desire outside assistance.[12] If these conditions are not met, it is quite likely that attempts at intervention will cause more harm than good and so will be unjustified.

There are, in sum, three categories of just cause for war: response to aggression, pre-emptive strike against likely aggression, and response to threats to the lives and well-being of citizens of other nations. Simply labeling a military operation 'an act of aggression' is, therefore, not sufficient to demonstrate that one has just cause for going to war. This is so even if the term is applied in accordance with our definition. A given military action may well be wrong, for example, but going to war in response will not be justified unless doing so will save substantial numbers of human lives. Or, going to war to protect the lives and well-being of citizens in other nations will not be justified unless the gain in human life outweighs the cost. We must guard against the temptation to simply use these categories as labels and then claim that war is justified whenever the label is applied correctly. In particular, when considering whether any of these three categories actually justifies going to war, two other classes of factors must be taken into account. One is proportionality and the other is the long-term and wide-ranging consequences of initiating war.

The doctrine of proportionality is simply that the means used must be justified in terms of the expected gain. This implies, for example, that a military response even to a clear act of aggression may be unjustified if the cost exceeds the expected gain. The wrongful theft of a small piece of uninhabited territory will not be just cause for a military response if it is likely that a major war with an aggressive military power will be in the offing. Or, in the case of the Khmer Rouge in Cambodia, it was argued that a military effort to save the citizens of Cambodia would have been extremely difficult to mount and quite unlikely to succeed. Of course, considerations of this sort rule out only a military response leading to war. Even if military action in Cambodia were not feasible, there was much else, short of war, that could and should have been undertaken to aid the Cambodians. Such measures might include economic boycott and diplomatic maneuvering in the UN, but they could also include positive incentives, such as granting aid in return for better conduct. We should not, in any case, underestimate the importance of publicity and repeated expressions of concern by national leaders.

It is more difficult to determine whether a given war may upset regional or global stability or strengthen the position of an aggressive and greedy military power. An otherwise justifiable war between two small nations may be thoroughly wrong if the conflict is highly likely to involve the great powers of the world and perhaps lead to nuclear exchanges. An otherwise justifiable conflict between, say, Israel and

Syria may not be if the United States and the USSR are likely to become heavily involved. Or, many have argued that it would have been a mistake to aid Ayatollah Khomeini's Iran in its war with President Hussein's Iraq even though the intervention could have been accomplished at small cost and Iran was clearly wronged by Hussein's invasion. The reasoning is that a clear victory for Iran in this conflict would give strong impetus to Islamic revolutionaries in other Muslim nations in the Middle East which could, in turn, seriously undermine regional stability, as well as unleash tremendous social upheaval and untold human suffering. If such arguments are successfully applied in these situations, they seriously undermine the case for what otherwise might have been fully justified warfare. Because human lives and well-being are at stake, these are ultimately moral arguments. In particular instances, such as the Iraq/Iran conflict, they may be sufficiently forceful to outweigh other possible grounds of justification.

A final, complicating, factor is that warfare may sometimes have just cause where there has been no previous military activity of any sort. David Luban has pointed to instances where a nation is in grave danger from famine, flood, or other natural disaster – or even through its own mismanagement.[13] Where large numbers of lives are at stake, he argues that other nations are obligated to help. If they fail to do so, Luban also argues that the needy country is justified in going to war to gain adequate supplies. Obviously, a whole host of moral issues are raised here, including the importance of national boundaries and the nature of the moral obligation to assist others – as well as questions about the practical likelihood that such military action will be capable of achieving its intended goal. Yet it is clearly possible that cases of the sort Luban envisages could arise and just as possible that in some of them warfare could be a morally justifiable response. This is so particularly if we keep in mind the principles that the interests of all human beings are morally significant and there is moral justification for having a system where the governments of nation-states give highest priority to the vital needs of their own citizens.

There is thus no simple test for determining that a given war is justified. Most often a war will be just if it is either a response to aggression, a pre-emptive strike against likely aggression, or an attempt to prevent great suffering of those in other nations. But these are only general guidelines. As we have argued, even clear-cut aggression will not justify war if more human suffering will result from fighting than not fighting. Each case must be decided finally by reference to proportionality and the long-term effects of going to war.

CHAPTER 6
Role of third parties

I Bystander status

The active participants in wars are not the only ones who must face the issues of just war. Bystander nations must also determine whether they are obligated to become involved in ongoing wars and whether they may justly be coerced into involvement. In some ways, the moral problems facing bystanders are even more subtle and complex than those of the active participants. The complexity ranges across several dimensions, including the status, rights, and obligations of bystanders, as well as the rights and obligations which belligerent nations have with regard to bystander nations.

It is surprisingly difficult to gain a clear definition of what a bystander is, as distinguished from an active participant in war. Nations that are not the main participants may be involved in wars in all sorts of ways. Cambodia, for example, was technically neutral during the period of United States involvement in Vietnam yet it was involved in the war in a variety of ways, including fighting with North Vietnamese soldiers.[1] In contrast, Saudi Arabia has technically been in a state of war with Israel for years. Yet it sent only token forces to participate in the 1973 war, and gives every indication of wishing to avoid direct conflict in the future.[2] Actual involvement in fighting by itself is not sufficient to distinguish between belligerent and bystander. So, 'bystander' is not adequately characterized as simply 'noncombatant'.

Neither is 'neutral' adequate as a definition of 'bystander'. 'Neutral', as it is used in international law, is defined in exquisite care and detail.[3] The provisions are complex, but the essence of the idea can be stated simply. A neutral nation is one that both declares an intention to remain uninvolved in combat and pledges to carry on its relations with belligerent nations in exactly the same fashion as before the war began. It may keep the same patterns of trade, cultural interchange, etc. as existed before the war. Though neutrals must remain strictly evenhanded in this sense, patterns of trade and other contacts will usually, in fact, be of greater assistance to one side than the other.[4] In return for exercising restraint, neutrals gain the right in international law to remain free of hostile acts by belligerent powers and also to remain free from coercive efforts to gain their active participation in the war.

Some writers appear to assume that the only option nations have is either to participate fully in wars or to adopt the status of a neutral.[5] It is important to understand that this is not so. Neutral is only one type of bystander. A nation, for example, may remain scrupulously non-combatant, but also be quite partisan. In the war between Iraq and Iran in the early 1980s France was a non-combatant yet wished to tilt the balance of the war in favor of Iraq. It did so by selling it sophisticated fighter planes and missiles. It was a non-combatant but clearly partisan, and its partisanship made a difference, given the shortages of equipment and material that plagued both sides. Partisan bystander status is by no means uncommon. The great powers practice it as a matter of course, given their tendency to stratify nations and events into the rigid categories of ally or enemy.

Control of arms sales or willing participation in embargoes on such transactions is the most direct form of partisanship and the sort which is likely to be viewed as one short step away from active participation in war. For this reason, and because such partisanship can often be highly effective in changing the course of the war, it is most likely to provoke a hostile response from one of the belligerents. France discovered this shortly after her arms sales during the summer of 1983 when, in the fall, she suffered a number of terrorist bombing attacks widely believed to have been initiated by Iran.[6] The United States also engaged in this kind of tightrope walking in its very extensive participation in the guerrilla war in El Salvador in the 1980s.[7] It got away with this without provoking hostile response because of the great disparity between the resources which it commanded and those of the hapless guerrillas. The United States was in position to make a substantial difference in the course of a war and was able to do so without becoming involved as an active combatant and without subjecting itself to any great degree of danger.

This last point is of considerable importance. One corollary of the view that war and strict neutrality are the only options nations have is that the only alternative to the strict restraint of neutrality is the full hazard of war. As the example of the United States in El Salvador indicates, nations can in fact often stray quite far from the legal requirements of neutrality without subjecting themselves to the devastation of war. This has central consequences not only for what nations may wish to do, but for what they may be morally obligated or even be compelled to do.

There are other classes of partisan activity less dramatic and less hazardous than arms sales, but consequential nonetheless. Nations can initiate or participate in economic embargoes of products other than arms. They can form, uphold, or abandon alliances with nations which may commit them to support short of actual combat. They can engage in diplomatic maneuvering by making speeches, participating in international investigations, issuing reports, posturing, and making displays of military force. These may not have so clear-cut an effect as arms sales, but will have consequences not readily ignored by any but the most hard-shelled belligerent.

Nations may, in addition, encourage or simply allow their citizens, either as individuals or as members of various sorts of groups, to campaign actively on behalf of a belligerent. International law allows citizens this partisan conduct. This seems misguided for several reasons. One assumes that the ultimate justification for rules about neutrality is that they protect the lives and well-being of ordinary citizens. If it is finally citizens themselves who enjoy these benefits, it is reasonable to require them to undertake some of the responsibility for maintaining the status of the nation as neutral. Doing so allows them to be active, that is, to play a direct role in maintaining national security. It would be mistaken to claim that nations cannot control their citizens to this degree since nations can and do control the lives of their citizens in various other ways. If it is argued that such constraints would infringe citizens' rights of freedom of speech, we note that the justification for assuming the role of neutral is to preserve national security and that such grounds are those most widely accepted for abridging this freedom. Further, what needs to be curtailed is not simply voicing opinions or criticizing governmental policy but actively organizing, campaigning, raising funds and providing material support for the belligerents. These fall more closely into the less rigorously protected areas of conduct than the carefully screened area of speech.

But there is a more basic mistake in the stance of international law on neutrality. It presumes both that states are the only actors in the international arena and that private citizens, unconnected with governments, can do little to affect the course of international events. Both presumptions are false. States gain whatever legitimacy and authority they have in international affairs by acting as the representatives of individual persons. Ultimately, all of their decisions and policies must be justified by their impact on the lives of particular persons. It is individuals, not states, that count in the final analysis. So international relations can and should make explicit reference to the legal rights and responsibilities of individuals whenever possible. Acknowledging their responsibility in maintaining neutrality is one way of doing this. An additional side benefit is that such provisions may aid in reminding states that they are not the only actors on the scene. However, this would count for little if private citizens could do little. The example of Jewish people in the United States supporting Israeli war efforts and in shaping United States and others' foreign policies should be sufficient to undermine this idea. Citizens can organize to provide material support. They can lobby. They can keep governments focused on issues of their concern. They can make contact with citizens in other countries. They can mobilize religious, educational, scientific, and business organizations on behalf of their cause. They can, in short, do a great deal, and their efforts can have significant impact. Because they can, they need to be assigned responsibility in international law. Present international law is deficient in this respect and should be amended to account for the importance of private citizens.

As it is elaborated in international law, neutrality is a passive condition. Neutrals, in effect, attempt to carry on their usual relations with belligerents as though the war did not exist, and the active participants in turn are obligated to avoid hostile actions toward them. But there is also another sort of neutrality, what we might call active neutrality. Active neutrality entails making positive efforts to be even-handed, along with attempts to maneuver oneself into the position of serving as an honest broker between the warring parties. The point of active neutrality is to put oneself in a position to do some good either by way of ending the fighting or mitigating its ill consequences. A nation may attempt to position itself as an honest broker by calibrating its trade policies so that they are equally advantageous to both sides. Or, it may maneuver diplomatically so that it is seen as equally friendly to both parties to the war. The policy of active neutrality has been employed by the United States on several notable occasions. The policy worked quite well when Henry Kissinger, and later Jimmy Carter, labored to end hostilities between Egypt and Israel. It was less successful when Alexander Haig attempted to do much the same thing during the war between England and Argentina over the Falkland Islands.

Old fashioned neutrality is thus just one form of bystander status during war. The other interesting and useful forms include active partisanship of one sort or another and the active neutrality mentioned above. In fact, explicit declaration of legally neutral status or lack of it may tell us very little of the actual role a nation plays in a particular war. It is perhaps most useful to simply and arbitrarily consider a nation a bystander if it is (a) not involved in combat and (b) has declared its firm intention to remain uninvolved. Having done this, the bystander is then faced with the delicate problem of attempting to decide which of the varieties of bystander status are practically and morally required.

II Arguments for and against

It may appear surprising that a nation should find it difficult, from either a moral or a prudential perspective, to justify keeping itself out of war. National leaders have an obligation to avoid war which is even stronger than that of the individual person to avoid conflict, and not only because the numbers of people involved are much greater. War is much more serious than personal combat. Not only is the range of its destructiveness much greater, but its consequences are extremely unpredictable, and its long-term effects may be enormous. Conflicts thought to be limited and manageable all too often evolve into affairs of much greater scope, as the United States found out during the course of its involvement in Vietnam. Wise leaders will feel an obligation to their people to avoid such uncertainties. In addition, a person who determines that he has a duty to accept a risk on behalf of another is deciding for himself and is suffering the consequences for himself. National leaders, in contrast, are deciding for others, who may have no say in the decision and who may suffer greatly because of it. Further, the individual who acts to aid another can normally expect to receive support from the social system in which he lives. There is no such effective system on the international level. The nation that acts for the benefit of others, particularly at some risk to itself, can usually expect to act alone. Lastly, it may be argued, as Michael Walzer has, that individuals develop intimate ties and obligations to one another as the result of functioning as part of a single community. Nations exist as part of no such structure.[8]

Nations, then, have a strong obligation to avoid going to war, one that may be overriden only by the most weighty considerations. In the case of bystander nations, in particular, war is avoidable in the senses that bystanders are not directly under attack and they are not party to the issues which spawned the war in question. Having no self-interest served in going to war and not having war thrust upon them, it may be worthwhile for a bystander nation to pay even a substantial price to avoid involvement. In these situations the price may be a fairly modest one of not overtly favoring one side or the other. But the price required may be larger. It may be the price of allowing the triumph of a wicked and greedy regime or a decrease in world stability. The price may involve paying an indemnity, allowing its territory to be used as a conduit for troops or material, or even giving up a chunk of territory. The price here is loss or infringement of sovereignty. Or, it may pay a moral price of standing by while human rights are violated in wholesale fashion. The price may indeed be so high that it may not be worth paying.

There are a number of ways of arguing against the usually sound view that war should be avoided. An especially strong argument is that the bystander nation may be trading a short-term gain for long-term disadvantage. The imperialistic nation that it mollifies today may be more ambitious a short time later. Or, it may be that the war will serve to create increased world tension and undermine stability in ways that will envelop the bystander in the future. Walzer points out that there is a strong moral as well as prudential dimension to these arguments.[9] By allowing others to fight the battles and suffer, even if they can successfully defeat the imperialist or restore world harmony, the bystander is in effect allowing others to sacrifice themselves for its own interests. Avoiding others' quarrels, Walzer points out, is one thing, but allowing others to suffer on one's own behalf is quite another.

It is also quite plausible to believe that nations, as do individuals, have a strong obligation to help others in need. Where other people are threatened with mass murder or mass slavery, the world community has a strong obligation to come to their assistance, even at the cost of going to war. Nations, as well, have an obligation to prevent the strong from preying on the weak and to prevent grave injustice from being done. It is possible, as Walzer argues, that those within a political community have a stronger obligation to look after one another than they do to assist those outside.[10] This view is not without difficulty, and there are philosophers who would not accept it. But even Walzer would agree that *some* obligation must remain, since much of what we owe to others is owed simply because of the duty to relieve suffering wherever it is found and not because of special ties which we may have to them. It may well be true that nations have the obligation to preserve themselves first and to look out for others later. But this does not imply that there are no occasions where nations will have some obligation to make sacrifices for others.

Sometimes, in addition, bystander nations will themselves be responsible in some ways for the outbreak of wars or the scope of devastation caused by them. Nations may encourage war by selling others arms or supplying them with advisers or expertise. Nations may sow discord and hostility among others, then stand aside when conflict breaks out. At other times nations will pursue trade or economic policies which bring others into conflict with one another. In these cases, where bystanders have some responsibility for the outbreak of war, they also have some responsibility to attempt to rectify the situation – perhaps even to the extent of becoming active participants.

Another, sometimes persuasive, ground for involvement in war is enlistment in a great and noble cause. Woodrow Wilson pictured United States entry into World War I as a moral crusade. Often the rhetoric used by national leaders to inflame their countries, and perhaps themselves, is enveloped in the service of the great cause. Strangely, this is both the most emotionally compelling and least intellectually persuasive of the classes of argument. It *is* the sort of argument that is likely to rouse citizens to action. They will enlist in wars in order to 'make the world safe for democracy' or to create peace and harmony. Aside from the fact that it is questionable whether any war has actually effectively served these causes, it is not clear, as we argued in the previous chapter, that they are compelling enough to justify going to war. It may well be important to make the world safe for democracy or achieve Franklin Roosevelt's Four Freedoms, but it is not obvious that they are sufficiently compelling to justify war.

Thus, nations in general have strong prima facie reasons for wishing to avoid war. But, there may also be instances where compelling arguments will override these. In order to make these decisions, nations must sift factors from a variety of categories. The kind of response they make to a war involving others must be the result of a complex calculation involving the seriousness of what is at stake for themselves and for others, the extent to which they can hope to cause effective change in a constructive direction, the extent of their vulnerabilities, and the extent to which a given war is likely to involve great hazard for them. With this information in hand, they may determine whether to become active participants, at one extreme, or strict neutrals, at the other; or adopt one of the options in between, ranging from active non-combatant partisan to passive partisan to active neutral.

III Choice of options

Surprisingly, the choice of full neutrality as defined by international law is justified in only a narrow range of conditions. This surprise may erode somewhat once we recognize that legal neutrality is practiced rarely in wars of any consequence by nations who have some stake in them. Partly, of course, this is because they can usually get away with partisanship or active neutrality without any great harm to themselves, but it is also in part because they often have sound reasons for becoming involved. Great Britain must be involved in any war involving Commonwealth countries because of its ties with them. The United States must become involved in conflicts involving its allies.

Legal neutrality is justified only if

1 there is no moral or practical reason to choose among the belligerents;
2 the bystander nation is highly vulnerable to violent responses from the active participants; or
3 there is little reason to believe that the bystander nation will be able to make any effective difference in the outcome of the conflict.

However, nations are not always justified in remaining indifferent to injustice or violations of rights that occur beyond their borders. As we have pointed out earlier, they may plausibly be thought justified in securing the well-being of their own citizens before those in other countries, but this does not imply that moral obligations do not reach beyond their borders. Other human beings have some moral claim on us simply because they can suffer. Or, on the practical side, nations would be foolish to attempt to remain aloof where they have a significant self-interest at stake or where some greater good, such as increased world stability, can be achieved by their involvement.

Even in cases where there is nothing to choose between belligerents, bystanders may have the obligation to become active neutrals and not merely legal neutrals if there is reason to believe that their good offices may significantly shorten a war or limit the scope of its destructiveness. From a broader perspective it may be important for nations to attempt to gain the status of active neutral simply so that the practice becomes a familiar one in the context of international relations. If nations become acclimatized to the efforts of others to be honest brokers when wars break out, thev may come to expect that such efforts will take place and prepare themselves for the possibility of engaging in negotiations. Also, this practice may set a useful example for other bystander nations who may be similarly situated to be helpful in other wars.

A nation may still have the right to legal neutrality if it is highly vulnerable to a violent response from a belligerent. Thus, if a bystander is likely to become actively involved in an all-out war if it is less than fully neutral, its attempts to remain strictly aloof are reasonable. If, however, the response which it can expect, though violent, falls short of all-out warfare, it may have the obligation to accept the hardship.

But, nations need not make sacrifices if there is no reason to believe they will bring any significant benefit. Even in this instance, however, it may be important to register disapproval of wrongful conduct and make some effort to bring it to a halt. It is reasonable to suppose that the world as a whole will be better off if nations clearly voice their disapproval of wrongful actions and support their concern with action. Even a futile gesture may sometimes be obligatory if it does not require a great price.

At times, however, bystander status of any sort may be unjustified. Nations have the obligation to become actively involved in wars when they are likely to do greater good than the harm which they will suffer. Wars, being what they are, this good is likely to be the negative sort of mitigating harm. That is, the good will involve stopping injustices, eliminating an outlaw regime or simply bringing an earlier end to the conflict. But the benefits may be positive also as where entry will contribute to greater harmony and stability or where it will substantially enhance respect for human beings. There are two qualifications on this general rule. For one thing, given the uncertainties and dangers of entry into war, the benefit sought should be clearly attainable and it should substantially outweigh whatever harms are expected. World events do not allow for the same precise calculations sometimes attainable in personal affairs or in highly structured situations. Given the uncertainty and pitfalls of expectation in this area – and the consequences of making a mistake – wise nations will only make a commitment where substantial benefits are clearly expected.

A second qualification is that nations cannot be expected to sacrifice themselves – any more than individuals can.[11] Further, national as against individual suicide is of much greater consequence not only because of the numbers involved but because a culture or way of community life may be at stake.

There are several general categories of cases where bystander status is unlikely to be justified. One is where a nation has the capacity to bring a long, sputtering, destructive war to a prompt conclusion at little hazard to itself. Another is where there are gross violations of the rights of individual human beings, as the murder of helpless populations. This is at least part of the reason why the Allies were morally obligated to enter into World War II. Finally, it is difficult to justify remaining a bystander where world stability and harmony are clearly at stake. Even here though a small, highly vulnerable and weak nation may be justified in remaining a bystander. What is clear is that nations are not generally justified in standing idly by while the world goes up in flames.

IV Coercion by other nations

Demonstrating that a nation is obligated to enter into a war is not the same as showing that others are justified in exercising coercion upon it by violating its bystander status. It may be that such overriding is sometimes in order but it is essential to clearly understand the instances when this may be so.

There are two importantly distinct ways in which bystander status can be violated. First, a bystander may be subject to unprovoked and unjustified violent action. Bombing shipping or cities or overrunning borders may be unjustified, unprovoked attacks – wrong for much the same reason that any violent act is wrong. Acts of this sort may nonetheless be justified when citizens of a nation at war are literally fighting for their lives as individuals or as a community.[12] It is important to be clear about this. Such extraordinary action is not warranted simply to insure the survival of a particular government or regime. A government is not a human being and does not have the same claim to continue in existence as individual human beings do. Even good and humane governments cannot justify such measures to remain in power – unless the government to follow is likely to be one which will endanger the lives and well-being of individual citizens. Governments matter only to the extent that their effects on individual human beings matter. Neither are such measures justified as necessary to win a just or morally obligatory war. A war that a nation is obligated to fight is not necessarily one that it is morally obligated to win – in a sense which would justify any measures, however cruel or inhumane, which might be needed to do so. Perhaps it is morally necessary to win some wars, ones in which world peace and stability are at stake, for example, but the same cannot be said of all, even just, wars.

A second, distinct, way in which bystander status can be violated is through the use of coercion to enlist a nation to a cause. This may be done in a variety of ways. One way is to use threats, economic coercion, blockade, bullying to wear down the bystander. Another way is to move troops into a nation, perhaps even to occupy portions of it, in such a way as to invite attack by the opposing power.

This sort of coercion, it may be argued, is wrong for a variety of reasons. For one, since wars are bad, it is wrong to dragoon others into them. Nations that have avoided war, and perhaps even made great efforts to do so, are greatly harmed if they are nonetheless compelled to become involved. For another, it is all too likely that belligerent powers are acting from self-interested motives in enlisting others.[13] They are, after all, at war, and war is always perceived to be desperate. So in their desperation they attempt to enlist help wherever they are able – apart from whether the bystander nation, itself, has compelling or even good reasons to join in. A third reason is that nations strongly value making free choices. They are always harmed when these choices are overridden, particularly in instances as important as those pertaining to war.[14]

Nonetheless, these considerations may be overriden under

genuinely extraordinary circumstances. As previously stated, a nation fighting for its life may be able to justify coercing another nation into going to war. This coercion is not warranted if the other nation will be sacrificed in the process of saving the coercing nation. But it may be warranted in demanding that the other accept substantial sacrifices, including going to war, to help. Of course, such action is not justified as a last futile gesture. Such sacrifice is only mandated if it is genuinely likely to make the difference between life and death.

The second sort of extraordinary circumstances are found where an outlaw nation is a genuine threat to the stability of the world community. Just as there are events larger than individuals, so there are events larger than nations. Where the future of the world is at stake, nations may legitimately be drafted to serve the cause, provided there is good reason to believe they will be of assistance in doing so. It is, of course, difficult to know when the stability of the world is at stake. It is also the case that nations will sometimes be tempted to misuse such arguments to serve their own selfish purposes. But we should not throw out good arguments simply because people will be tempted to put them to bad use. Genuinely catastrophic cases do arise, as when the world faced the Axis powers. There are extraordinary cases and extraordinary measures are required for dealing with them. Just as the free choices of individuals may be overriden to deal with special cases, so the free choices of the governments of nations may be sometimes overriden.

V Conclusion

The situation of bystander nations is by no means simple. Bystanders often have important responsibilities to assist those who are at war. Taking refuge in legal neutrality is appropriate only in a narrow, and well-defined, array of circumstances. In some, nations may be obligated to forgo bystander status and become active participants in wars. In even more extraordinary circumstances, they may be compelled to do so by others.

Bystander nations will be tempted to ignore their responsibilities when war comes. This is understandable both because of the sacrifices that may be expected of them and because, historically, nations have all too often used wars to serve their own self-interest and greed rather than carry out their responsibilities. Nonetheless, if nations are to work to fashion the world into a more humane place, one where wars are infrequent but fought for good reasons when necessary, then they must examine their duties as carefully when they are not at war as when they are actively engaged.

PART THREE:
ISSUES OF FIGHTING WAR

CHAPTER 7
The enemy

I Harming the helpless

Three Khaki soldiers approach an abandoned village from the North just as three Brown soldiers approach from the South. They surprise each other as they meet in the center of the village. A fire-fight ensues and it ends when all six are dead. One has been shot, another crushed by falling debris, another burned by a gasoline fire, another stabbed, still another dismembered by an explosion, while the last has been asphyxiated. This is a realist-like, 'War is Hell', portrait. Were a squad of Browns to sneak up on some Khakis and destroy them 'before they knew what hit them' that too would be a realist-like portrait. It would be the same had the Khakis led the Browns into a trap by cleverly changing some road signs. Dwelling on these and similar portraits, the realist would claim that ethics has no place in them since the prohibitions we normally associated with ethics such as those concerned with killing, taking unfair advantage of someone and lying have been cancelled. It is true, as we have indicated in the Introduction, that these portraits could be redescribed in ethical terms. It could be said that once soldiers put on their uniforms and participate in a war, they are still acting under moral rules but ones that are different from those in peacetime. The Iron Rule of war, compared to the Golden Rule of peace, might thus be stated as 'Do unto others (your enemies) as they do unto you.' Variations on this rule might be 'do it unto others before they do it unto you' or, worst yet, 'Do unto others what they cannot do unto you.'

Now these rules strike us more as parodies of moral rules than moral rules as such. But notice that in war they can be applied by both sides. If these were the moral rules of war, both sides would be governed by them. That would seem to leave a sense of fairness to war and, thereby, to leave some ethics in war after all. However, one need not accept this sophistical kind of argument to find ethics in situations like those portrayed above. Even if we grant that the Iron Rule and its variations are not moral in nature, there are still many seemingly moral questions having to do with how the battle is fought and what happens after it is over that need to be answered.

Concerning the former kind, we intuitively sense that it would have been wrong for the Browns to kill the Khakis had they both been able to surprise and easily capture them. In addition, once they are captured, we want to say that the battle is over and the old ethical rules come into play again. Our intuitions tell us that killing captured prisoners, when these prisoners could just as easily have been returned to the Brown lines, is murder.

But the question is: Are our intuitions fooling us here? What, to return to Hare's distinction between intuitive and critical thinking, does the latter form of thinking tell us about these capture situations? If, as Hare does, we count preferences on that level, the moral thing to do is clearly to take prisoners and not to shoot when shooting is not necessary. The Khakis would most probably prefer to be disarmed by capture than by destruction. As to the Browns, since their preferences are to disarm the enemy, and since it should make little difference to them (in most situations at least) what the method of disarming is, the Khaki preference not to be shot should override whatever small tendencies the Browns might have to shoot them. Actually it should make little difference here whether or not a utilitarian approach like Hare's is adopted in doing critical thinking. If, on the critical level, the final appeal is to rights, it should also be rather easy to conclude that, if avoidable, no shooting should take place. If we assume that all humans have a right to live or that they all should be treated with equal concern and respect, needless killing would be a violation of rights. Roughly, the same thing could be said if the critical thinking were duty based. We have, the duty-based theorist would say, a duty to respect life. If the Browns can disarm the Khakis without killing them, then they have a duty to do so.

Other easy cases can be cited. A Khaki patrol operating behind enemy lines spots three ambulances heading away from the front lines. The patrol has set up a trap with machine guns and other weapons. As the ambulances approach it fires and destroys the vehicles and all those in them. Surely, intuitively, we would condemn such action. But the question is what should be said after examining the Khaki attack critically? In Hare's terminology, would critical thinking lead us to change our intuitions or leave them as they are? Taking account of the preferences of both sides leads to the same conclusion: viz., that our intuitions should be left alone. From the Khaki point of view their preferences are not being satisfied by their attack. In one sense they are doing what they want at that moment. But their considered preference is to act in ways that will facilitate their war effort, and this particular action is doing very little along those lines. Killing soldiers in the Brown army who have already been disabled hardly represents good use of their time and equipment. By shooting at ambulances, they even risk having themselves spotted behind enemy lines 'all for nothing'. On the other side, the preferences are such that even when well, the Brown soldiers would prefer in a sense that the Khakis not shoot at them. But the Brown soldiers are participating in a war and in a more general sense their preferences are, when healthy, that they take part in it. But when they are wounded, their preferences have likely changed. It is as if, when wounded, they have retired from military service and become civilians again. It could actually be argued that the wounded soldier is less dangerous to the enemy than a civilian since the latter could quickly don a uniform and fight in battle, whereas the wounded soldier cannot. So if, as we have been arguing, everybody's preferences in a moral situation need to be taken into account, the strong preferences of the wounded soldier not to be shot at plus the considered preferences of the attackers not to waste ammunition and effort on worthless targets, make it clear that the attack on the ambulances is simply not the thing to do morally.

Much the same preference analysis can be given to a situation where the wounded are still lying on the field of battle. Little is to be gained if a soldier finishes off a severely wounded enemy soldier just as the medics are arriving to clean up the battlefield. The wounded soldier's preferences, were he able to express them, would (normally) be that he be saved rather than finished off. From the attacking soldier's point of view, his preferences are not satisfied by the attack. As a soldier representing one side of the war, he gains very little, while the other side loses a lot. This might still seem like an argument for encouraging attacks on wounded soldiers. However, in this case, what the other loses is not military capability but preference (satisfaction) of those who have already been put out of commission militarily.

II Surrendering

Other easy cases can be imagined, including ones having to do with abusing prisoners. But there will also be a variety of hard cases to deal with, and it is to these cases that we now turn to see what can be said about them. Certainly many hard cases will involve civilians and what soldiers can and cannot do to them morally. However, for now, and for the sake of keeping the scenarios as simple as possible, we will assume that there are no civilians around the battle area, and also assume that our problems are on the level of soldier versus soldier rather than on a higher level. Many of the problems on this micro-level will have to do with surrender. For instance, Hastings and Jenkins describe an incident near Goose Green in the Falklands War as follows.

> While B Company swung around the airfield in a wide hook to approach Goose Green from the southwest, C and D Companies linked for a combined assault on the schoolhouse. The Argentine defenders fought back fiercely until a white flag suddenly appeared from an enemy position. One of D's subalterns, Jim Barry, moved forward to accept the surrender. He was instantly shot dead. It was almost certainly a mishap in the fog of war rather than a deliberate act of treachery, but the infuriated paras unleashed 66mm rockets, Carl Gustav rounds and machine-gun fire into the building. It was quickly ablaze. No enemy survivers emerged.[1]

One way we sense that we are in trouble in dealing with these kinds of cases is that unlike the easy ones, our intuitions do not necessarily speak to us with one clear voice. Perhaps some people are shocked when they hear that the British fired with such vengeance at those who seemed, in spite of the Jim Barry shooting, to be in the process of surrendering; while others would have been surprised had the British not fired. But most of us will have a sense of conflict since our intuitions pull us in opposite directions. At times we will feel that the British acted wrongly and at other times not. There is no question of the wrongness of feigning surrender, if that is what the Argentine soldiers were doing. Such behavior undermines the surrender convention by making it far more difficult for those who, at other times and at other places, have every reason for surrendering to do so. The more difficult question is how are those who are taking prisoners supposed to react to an enemy when some of them, acting pretty much on their own, are playing surrender tricks on their would-be captors?

A look at the surrender convention helps to answer this question. Surrendering is a convention much like promising, apologizing and stating are. All of these practices are speech acts which, when not done in accordance with certain rules, become non-acts (i.e., are void).[2] If a person has done nothing wrong (either by way of commission or omission) to another, he can hardly apologize to him. The act of apology requires an unfortunate prior (i.e., preparatory) act to take effect. In a like fashion, certain rule violations void the act of surrender.

Speaking in the most general terms, the act of surrender falls under the class of declarations.[3] What is distinctive about declarations is that they make things happen as if by magic by their very performance. A request, a speech act falling under the heading of directives, hardly has this same magical power. A request to open a door does not automatically get the door opened for us. Nor does a promise, which is an example of a commissive, get the promised act done automatically. One may carry out the promise but then, depending upon a variety of things, he may not. But with a declaration such as 'You're fired' uttered by an employer to a (non-union) employee, the poor employee has lost his job as soon as the employer has finished speaking. As a declaration, the act of surrendering is also automatically complete and consummated in its very doing.

Related closely to this automatic feature is the act's unilateral status. Surrendering is not like betting which is an act or activity requiring the cooperation of at least two people to consummate.[4] It is true that some surrenders are negotiated so that they do not take effect until, like betting, both sides have agreed to a deal. It is also true that if the two sides are not clear what the surrender convention is as when a soldier does not understand his enemy's linguistic equivalent of 'I surrender' it, again, may take time to negotiate a surrender. Nonetheless, if an enemy soldier is clearly understood as surrendering, since surrendering is a unilaterial act, it is not up to his captors to reject his offer. To speak here of an offer of surrender is, therefore, strictly speaking, a misnomer. One cannot respond to the surrender by saying 'We don't accept your surrender' and then start shooting. Indeed, one may shoot after someone has surrendered and thereby do something immoral and no doubt illegal. But the reason something wrong has been done here is that those on one side have already surrendered by raising their hands and doing whatever else is part of the convention of surrendering.

The nature of surrendering as an automatic and unilateral act thus puts the moral burden on those taking prisoners. What they do, whether they shoot or not, is dependent upon what others do. However, they have no say provided, always, that the act of surrender is performed in the appropriate manner. It will not do for the 'surrendering' soldier to say publicly 'I surrender' with gun in hand and a finger on the trigger. Were he to be shot dead on the spot just after saying these words, no one would accuse his slayer of killing a prisoner. Nor could anyone be accused of failing to accept a surrender if he sees that an enemy soldier, pretending to surrender, is strapped with demolition charges. The failure to follow the rules of surrender voids the surrender attempt and so relieves those taking prisoners of any responsibility not to shoot.

It becomes more difficult to know what to do, however, when a platoon has lost several men as the result of some scattered false surrenders, and now is faced with the problem of a new batch of enemy soldiers who are apparently eager to surrender. In order to keep from taking further casualties, should they simply shoot anyone who looks like he is trying to surrender? Certainly shooting would be the safe thing to do. False surrenders would no longer cause any problems. Still, such shooting would result in killing many enemy soldiers who have legitimately surrendered and, as a result, killing soldiers whose preferences are not to fight. In and of itself, these considerations should encourage one to pause before pulling the trigger. Coupled with the fact that other alternatives to shooting everyone in sight are available, such action should not be condoned. It should be kept in mind here that the enemy's false surrenders are not the sole cause of the casualties. Those who take casualties have to be careless as well. If, as the enemy approach to surrender, they are asked to stay away at a safe distance, take some of their clothes off, raise their hands, turn completely around and do other such things, there is no reason for their captors to take further casualties. No doubt there will be times when all these procedures cannot be followed, and at these times there may be doubt as to whether the surrender is feigned or real. Or it may be that the surrendering soldier has an unfortunate twitch that looks for all the world like an attempt to reach for a weapon. In such cases it may be better to be safe rather than sorry. But the vast majority of surrender cases ought to be handled without shooting when, as we have been assuming, the false surrenders are occasional and random.

But how about those who are surrendering in the heat of battle? Should soldiers under fire cease the attack in order to take prisoners? If there is nothing technically wrong with the actual acts of surrender, it would seem that they ought to be accepted. The surrendering soldiers, let us assume, have laid down their arms and, at the moment at least, have little inclination to pick them up again. They have had quite enough of the war in which they find themselves immersed. Further, they have uttered the magic words and/or have indicated that they have surrendered in some other traditional way. However, let us also assume that by detaching some men so that they could round up prisoners, the attack that might have been successful had it been pressed fully will now not be. So the problem here is not like those we have been dealing with where those effecting a capture have nothing to lose by doing so. In these cases, the preferences of the two sides seem to be at odds with one another.

Surely even in the heat of battle most surrenders should be honored. Especially in large battles, a few can be assigned the task of rounding up those who are surrendering. Yet even if no one is available, a partial surrender can often be effected by doing such things as taking weapons and shoes away from those who have indicated that they have had enough of war. However, it is not these kinds of situations on which we are focusing, but rather on those where the battle is so heated that there is no time to effect even these partial surrenders. In these settings the choices are to:

1   pursue the attack by bypassing those who have surrendered (and thereby risk the possibility that the prisoners will later pick up their weapons again);
2   pursue the attack by shooting all enemy soldiers in sight even those who have indicated that they wish to surrender; and
3   stop the attack in order to take prisoners.

Of these three options, the first is certainly the least appealing since it seems to be rewarding the immoral soldiers and punishing the moral ones. If the attackers do not return for a while, those who have surrendered might just be tempted to pick up their arms again and shoot in the back those who have captured them.

So the main conflict this field problem in ethics poses is whether to choose options 2 or 3. Of these options, the latter fits our intuitions best. Since in the vast majority of the situations when soldiers put their hands up in surrender, they are taken prisoner, it would be in accord with our intuitions to accept all those who legitimately surrender as prisoners. Especially if there had been no time in advance to discuss difficult cases like this one, the overall rule about rules that we probably should follow is 'When in doubt, follow the (intuitive) rule.' Intuitions aside, however, a good case can be made for adopting option 3 on critical grounds as well. It could be argued that maintaining the convention of surrender is important in war. When exceptions are made which allow soldiers to shoot those who have surrendered in order to pursue the battle, that convention is jeopardized since one exception inevitably leads to another. Controlling the behavior of soldiers in war is especially difficult and it is best, given that fact, to ask them to follow exceptionless or almost exceptionless rules. Thus defenders of option 3 are telling us that even under the pressure of extreme military necessity, exceptions to the rule should not be allowed. That is the nub of their disagreement with those who would defend option 2.

In response, the latter group would grant that 'military necessity' is a sorely abused expression. No doubt many surrendering prisoners have been unjustifiably killed under its banner. But it does not follow, they would argue, that although the appeal to military necessity is commonly abused that, at times at least, it does not have moral merit. What if, as the result of breaking off the battle in order to take prisoners, the enemy is given time to regroup? What if, instead of a quick defeat had the attack been pressed, a second attack must now be mounted? What if, as the result of the second attack, the enemy is defeated but at a much higher cost of life to both sides – much higher even after taking account of the surrendered soldiers' lives? Militarily, their argument is that there is no question what should have been done. Had the officers and soldiers involved in the battle pressed the original attack and won the battle more quickly and cheaply, they would have done the right thing. Morally, the defenders of 2 would add that the same conclusion follows. So long as the cost in lives (on both sides) is less by pressing the attack, it at least is not obvious that it would be morally wrong to do so.

III Tolerance

The arguments on both sides can be summarized by noting that whereas the defenders of option 3 point to the disutility of violating a rule, while paying less attention to the costs in lives of the battle itself, the defenders of option 2 point to the disutility of the loss of life, while paying less attention to how the rule about taking prisoners is eroded. Put just this way, the pull in the arguments from both sides still seems relatively equal, even though we have moved the argument from the intuitive to the critical level. Indeed the burden of our argument will be that, in cases like the one we have been discussing, there may be no right or best answer that critical thinking can deliver to us. One way to appreciate this point is to imagine two scenes following two separate battles. In the first, a captain is brought before a hearing because he has been accused of ordering the killing of enemy soldiers just as they were surrendering, but just after the battle had ceased. In the second, a captain is brought before a similar hearing because he has been accused of ordering the killing of enemy soldiers just as they were surrendering, but while the battle was still in progress. In both cases, if there were no disputing the facts that the killings had taken place, we would suggest that the hearings should arrive at two different decisions. The first should lead to a court martial and conviction, the second to a different judgment – one that requires the introduction of the concept of tolerance to be understood.

Tolerance is not the happiest of concepts. Although in a liberal society tolerant people are praised, they are also pitied to some extent. This is not just because it is difficult to become a tolerant person, although that is part of it. Being honest or brave is not easy either. Rather, tolerant people are pitied because in being the way they are they 'have to bite their lips' as a part of the process of putting up with certain conditions in the world. Tolerant people do not have the luxury of feeling good about what they are doing. Being tolerant has a bitter-sweet taste to it a little like what a child experiences when, after a nasty fall, it is praised for not crying.

The bitter-sweet taste of tolerance and its cognate concepts is accounted for by its function of sitting between extremes. Tolerating involves having a negative attitude towards some person, action or condition but, at the same time, not acting as this disapproval normally directs us to act because of some further consideration. An example here would be a captain's failure to court-martial his sergeant for stealing from prisoners because of the sergeant's great leadership abilities. It is this combination of *disapproval in attitude* and *restraint in action for a reason* that makes tolerating so uncomfortable.

This combination is also what makes it such a unique and interesting concept. It sits uneasily in the middle between intolerance and acceptance. It might be thought that this trichotomous feature of tolerance is not unique since indifference represents an option between good and bad. Likewise right and wrong can be trichotomized clumsily in terms of neither right nor wrong. But tolerance is trichotomous in a different way from these concepts because 'indifference' and 'neither right nor wrong' lie outside the realm of moral concern. In contrast, the restraint of tolerance can be within that realm. Often, of course we tolerate for non-moral reasons as when a captain tolerates his superior's stupid behavior because he wants a good letter of recommendation from him. But sometimes we tolerate for moral reasons as when we put up with the silly opinions of political commentators because they have the right to freedom of speech.

Some forms of tolerance have another interesting feature. We do not always let those we tolerate go their own way. We watch them to see whether they will slip over the line into intolerable behavior. In effect, we are telling them to be careful. This warning or threatening feature of the tolerance concept is important for our purposes since it is helpful in dealing with the notion of military necessity. Returning to the two captains, the first one, the one who ordered surrendering enemy soldiers shot after the battle was over, cannot justifiably appeal to military necessity to save himself from a negative verdict at his court martial. It might have been militarily inconvenient to take prisoners at that time, but hardly militarily necessary. The question of how to respond to the second captain is another matter. By ordering the fighting to continue, a battle was won that, very likely, would not have been. 'It was,' as he might have said in his report, 'necessary to do what we did.' But we should be clear about the exact nature of the necessity involved here. He could have meant that if he had not acted as he did, the war would have been lost. In the kind of small battle with which we are concerned, it is not likely that he meant this. More likely he meant that the battle would have been lost had he not acted as he did. But let us assume that even this sense of military necessity was not what he meant since he and his superiors knew that, in the end, his unit was stronger than the enemy's. All that he meant is simply that on military grounds what he did allowed him to win a quick and cheap victory. It was necessary, if a quick and cheap victory were to follow, for him to give the orders to continue the attack and not take prisoners.

He might have thought that he was right morally as well in acting as he did. After all, military and moral necessity are not incompatible, only different. In justifying his action had the captain said 'I had to give the order in order to keep my losses at a minimum' he could have been speaking either militarily or morally or, perhaps, both. In the narrow sense of giving a moral reason, what he said cannot be such a reason since 'my' suggests that he is taking account of the preferences of only his side in the battle. Yet, in a broader sense, he might be claiming to have acted out of respect for the lives of his people and, possibly, for the lives of the enemy also. In any case, let us assume that he thought he was acting morally.

Others will have their doubts, and one reason they might is that he has not shown the proper respect for the surrender convention. His actions, they might think, erode the convention even if they do save lives in this particular battle. At least they might wonder about this as they balance the lives lost in this battle against the erosion of a convention that will likely cost other lives in other battles. Their wondering is particularly appropriate since we are assuming that the captain had received little or no guidance from above as to what he ought to have done in this sort of situation. Given this fact, and given that the conflict between saving lives in that battle and saving the convention (and possibly lives in future battles) is difficult to resolve even on the critical level, it would seem that a tolerant response to what he did would be in order. Tolerance would also be the proper response to the actions of a third captain who, finding himself in the same kind of situation as the second one, thought more of the surrender convention than the casualties that might be incurred in the battle at hand and, as a result, delayed the attack in order to gather up some prisoners.

We do not mean to suggest that all serious moral conflict should be met with a tolerant response when it is raised to the critical level. In the case of the second and third captains they had neither the luxury of critical thinking nor direction from their superiors. They had to act on the spur of the moment on the intuitive level. However, when their superiors critically review their actions in the circumstances under which they were performed, including the absence of directions from above, our argument is that a tolerant response to them is perfectly in order. There is, of course, the separate question of whether headquarters will make policy for future situations once they complete their critical thinking. At that time, they may make policy in accord with what one but not the other captain did. But this additional thinking might just leave things as they were by turning what might otherwise have been thought of as a temporary tolerant response to the two captains into a principled tolerant response for all future captains. This may happen in war, as it does in life generally, because (moral) conflicts often yield judgments that are too close to call one way or the other. It may be that since wars are fought under the worst conditions, thereby making our judgments extremely difficult, the concept of tolerance is especially at home in war.

Those who like their moral decisions clean-cut, either right or wrong, may be upset by this conclusion. Referring once more to the slippery-slope argument, they will suppose that extensive appeals to tolerance concepts will erode our moral standards. One puts up with first this, that and that, and then finds himself tolerating everything and everyone (i.e., being intolerant of nothing and nobody). Although, surely, an appeal to tolerance has just this danger it also can have the opposite effect. With dichotomous thinking, the tendency is to merge what should not be tolerated with what should be tolerated. As a result what is called bad, not right or wrong, includes behavior and conditions that are not responded to in a consistent manner. Sometimes what is labelled wrong is fought vigorously, sometimes not. This response ambiguity can be mitigated by the use of the tolerance concept, leaving it clearer than before that what is not to be tolerated (i.e., what is forbidden or prohibited) means just that. Although some people will apply the concept indiscriminately by becoming overly tolerant, others will see that concept as helping them to firm up their convictions about what ought not to be done. Very likely it is not the concept itself that erodes standards but people's response to it. Those who are, as we say, morally weak probably will find it a dangerous concept to use. Yet for everyone, even the morally weak, it has a clarifying function that can be usefully employed in dealing with issues in military ethics.

Discussing two more cases should suffice to give a clearer sense of how the tolerance concept can be employed on both the intuitive and critical levels. The first is another surrender case. Two entrenched units within sight of one another are under attack from a powerful enemy force. Unit A has taken most of the punishment. It has taken many casualties and has little ammunition left. Its commander, seeing no sign of relief, decides to show the white flag. The enemy ceases firing and orders unit A's people to come out with their hands up. Unit B's commander is incensed at A's surrender and orders his troops to fire warning shots in A's direction. Unit A scurries back to its position but still keeps its white flag showing. The enemy now concentrates all of its fire on Unit B but, at the same time, goes to the trouble of signalling Unit A that it is honoring its surrender. As the battle between the enemy and Unit B continues, Unit A, to everyone's surprise, receives a limited supply of ammunition. At this point, A's commander pulls down the white flag and orders his troops to resume the fight. Not surprisingly, when both units are later overwhelmed by the enemy, A's commander is treated rather brusquely.

We will deal with only two of several questions posed by this case study: viz., should Unit A have resumed firing and, prior to that, should the enemy have resumed firing on Unit A after it had failed to effect the surrender? We will deal with the second question first.

It appears that the honorable enemy commander saw his problem as not being able to effect a valid surrender through no fault of his own. For him, the problem was one of clearing the battlefield of soldiers who have already taken themselves out of the battle. He would have preferred to do it immediately, but since his people dominated the battlefield, he figured that it would make little difference if he collected Unit A after he had disposed of Unit B or before. Although he miscalculated to some extent when Unit A was resupplied with ammunition, it is difficult to fault the enemy commander morally since he behaved humanely and in a way that sustained the surrender convention. However, had he resumed firing on Unit A, he would not have acted dishonorably (i.e., by doing something intolerable). In the heat of battle, when he had to act intuitively, he had to figure quickly on the possibility of an unforeseen turn of events that could endanger his men and a favorable outcome of the battle (from his point of view). In any case, as the surrender scene developed, it became clear that the two units were operating somewhat in tandem. He could have argued that since they are fighting together they must surrender together. Unit A's surrender need not, therefore, have been taken as a valid one. Had he responded by resuming the attack, his actions would not have been intolerable even if they would not have been totally acceptable either. Instead they would have been tolerable.

Unit A's commander was in a more precarious position morally. Assuming that he acted rightly in surrendering in the first place, he was in the embarrassing position of surrendering and reaping the benefits of a surrender by receiving a respite, on the one side, and resuming the fighting once he had been resupplied, on the other. In short, he was committed to fight on, but he was also committed not to. So his dilemma is genuine in that he was pulled in two different directions in trying to do the right thing. Whether judged by his own people or the winning enemy, once again, a judgment of tolerance is in order. Had others on either side put themselves in his situation (imaginatively), many would have felt guilty no matter what they would have done.

The second case we are considering is more straightforward. In this case, a unit is operating deep behind enemy lines in a jungle. The unit is a small one of ten men that, at most, could guard two or three prisoners. Well into the mission, but at least a week before returning home, it finds itself stalking an enemy column. As the unsuspecting column of thirty or forty men approaches, it is obvious that it is bedraggled. Most of the enemy is without weapons, it seems leaderless and there are many wounded among its ranks. It also seems lost. Putting it simply, the waiting commander has the choice of destroying, helping or ignoring the column.

If he chooses the first option it will be more a slaughter than a battle. The second option has some risk to it but not much since even if the enemy reaches a friendly base and reports the presence of a mobile enemy force behind its lines, the unit will be far from the scene where the two groups met. Whatever the risk, the kind of help it can give is minimal, although it could at least send the column off in the right direction. Finally, by ignoring it, the waiting commander could figure that these soldiers will probably perish in the jungle and thus never live to fight another day. After considering his options the commander lets the column pass and goes looking for worthier targets. It would seem that by not destroying the bedraggled column, the patrol commander is not carrying out orders since destroying the enemy is part of his mission. Presumably, however, he had a certain amount of discretion which he exercised. Now discretion is a next-of-kin concept to tolerance. To give someone discretion is to tell him in advance that a certain range of behavior options will be tolerated and/or accepted. Presumably, headquarters would tolerate the commander's decision not to attack the bedraggled column since that column was self-destructing, but it would not tolerate behavior that always avoided contact with the enemy or always attacked ambulances and hospitals.

Our problem, however, is not to assess the range of discretion allowed the commander by headquarters, but to assess morally whether what he did is tolerable, intolerable or acceptable. Let us assume that his patrol stalked the bedraggled column long enough so that he and those with him had time to engage in a considerable amount of critical thinking. Had they done so, part of their job would have involved placing themselves imaginatively in the position of those in the bedraggled column and asking 'What would they have wanted us to do?' A reasonable answer would be that they would have wanted assistance the most, and to be slaughtered the least. Beyond that it would be very difficult to know what their preferences might be. Once restored and directed back to their own lines, would they, after further rest, prefer to resume their role as fighting men? Or would they simply surrender when the next occasion arose? For their part, the preferences of the patrol are to not endanger their lives at the moment, and to not have to face these same enemy soldiers in the future. Ideally, in doing critical thinking here, account needs to be taken of the preferences of the commanders and military personnel on both sides as well as others, civilians included, who would have some preference about this incident were they to know about it. In practice the role-reversing of the mobile patrol commander would need to be limited to only those who would likely be affected directly by his actions – including those who would later be killed in other battles by those he had helped. Assuming, as has been suggested already, that he could not be certain how the enemy would behave if he helped them to return to their lines, he probably did the right thing in not helping them. Putting it differently, it would have been wrong (i.e., intolerable) for him to help the enemy in this situation.

It remains then to decide whether he should have ordered his men to shoot or let the column pass. The advantage of the former is the certainty involved. By letting it pass, probably those in the column would never fight again. But then one could never be sure. On the other side, there is little point militarily and morally in shooting. Ammunition could be saved for a more important battle. Further, very likely his men had no stomach for a slaughter. So in the end although the shooting option might be thought of as tolerable, it would seem that the let-them-pass option is the most preferable and that, therefore, the commander did the right thing. In this case there is no need to invoke the tolerance concept.

IV Retaliating

There was a tendency not to apply the universalizability principle to the local cases discussed in this chapter up until now. It is true that when one local commander's actions were tolerated, the similar actions of other commanders in similar situations were also tolerated. In that sense the universalizability principle was applied. However, by saying that their behavior was tolerated subject to later review, the implication is that behavior like that tolerated in the past might not be in the future. In dealing with local skirmishes where the unexpected might and often does occur, temporary tolerance is, it seems, the order of the day.

We now shift the focus of attention away from skirmishes to larger scenes where the key decisions are made at headquarters. With the shift, new issues arise that could not arise before, although all of the old issues such as those concerned with surrender remain – albeit in some different form. With the shift, as well, temporary tolerance recedes into the background since, as policy makers, those at headquarters have the time to do some critical thinking. Any failure on their part to engage in such thinking would be an inexcusable moral failure as it would not be for the local commanders who have to rely on their intuitions or some quick-and-dirty critical thinking in order to decide what to do.

As in the cases earlier in this chapter, a list of intuitively immoral orders can be assembled with little difficulty. The list below is concerned only with intuitively immoral acts toward the enemy military personnel. Obviously the list could be made longer by including immoral acts toward civilians, neutrals and one's own military personnel.

1 Any and all weapons are to be used to destroy the enemy.
2 Attack the enemy's medical personnel and facilities.
3 Attack the enemy's religious personnel and facilities.
4 Enemy wounded are to be finished off.
5 Enemy lifeboats are enemy vessels and those in and around them are the enemy. Both are to be treated accordingly.
6 Similarly, enemy descending by parachute (even from an airplane in distress) are still enemy and should be treated accordingly.
7 Do not take prisoners.
8 When, unavoidably, prisoners are taken, they are to be killed if they cannot be used to advantage in some meaningful way.

The discussion of the previous cases in this chapter suggests that these intuitively intolerable policies and orders will be assessed by critical thinking in the same way. Whether that critical thinking is based on goals (preferences), rights or duties will make little difference. Each basing mode, using its own special vocabulary and reasons, will conclude, for example, that killing off the enemy wounded is immoral. Each will also conclude that prisoners are not to be treated just as things. As usual, in real life and in the imagination of the theoretical ethicist, exceptions will be discovered to these rules so that it will be appropriate at times to finish off an enemy soldier. If nothing else, there will always be the case of the enemy soldier (or a friendly) who begs to be killed because of the terrible pain he is experiencing. Presumably, however, exceptions like these will be few in number and will not seriously erode the rule issued from command that, whenever humanly possible, enemy wounded will receive whatever medical treatment is available.

Although both our intuitions and critical thinking will affirm that the above list is immoral, there are two important problems that need to be faced with respect to it. The first is how a commander should respond when facing an enemy who is living by these rules. Is the proper response retaliation? Second, in relation to number 1 on the list, what weapons would a commander not use? These two questions are not completely separate, since the weapons used by one side may partly depend on those used by the other. Nonetheless, for the remainder of this chapter, we will deal with the first of these two questions only since the immoral acts on our list apply primarily to military personnel. The weapons-use question will be the topic of the next chapter, and will serve as a lead-in to a discussion in the chapter following that one as to how civilians should be treated by military forces during wartime.

Let us assume that at first there are scattered reports that the enemy is fighting a dirty war. In various skirmishes, wounded are reported shot, stabbed and clubbed to death. Pilots report enemy fighters shooting at their buddies who, in distress, are descending by parachute. This latter practice seems especially prevalent over friendly as against enemy skies. Also prisoners lucky enough to escape report that unhealthy prisoners are shot and the healthy ones worked to death. Stories like these continue to come in so that, after a time, it becomes apparent that they are not idiosyncratic but reflect enemy policy. What doubts remain are eventually dissipated when documents captured from enemy officers make it clear that they are under orders to fight a dirty war.

Certainly one inappropriate response to such immorality is to do nothing. At the very least it would be appropriate to take notice of these immoral acts publicly so that one's own people, one's allies and bystanders become aware of what is happening. At the other extreme, it would also seem intuitively inappropriate to retaliate by escalating the level of immorality. We will see shortly whether and how inappropriate such an escalated retaliatory response is on the critical level.

To retaliate is to do something nasty to the enemy after he has done something nasty to one's own people. But it is much more than that. A retaliatory act does not merely follow another act in time. An act becomes retaliatory only by being linked to an earlier act. Normally the linking is done by a third act, a communicative or speech act, that says 'This military act is in response to that one.' Without the communicative act which can precede, accompany or follow the actual retaliatory act, the enemy might miss the connection between the two acts of war. This is likely to happen if the retaliatory act is dissimilar to, comes much later than, or is geographically distant from the original immoral act. In most, but not necessarily all cases, labelling an act retaliatory comes down to issuing a threat of the form 'If you continue doing such and such immoral acts, we will do thus and so (also an immoral act).'[5] The 'such and such' and 'thus and so' need not be, although they usually are, the same kind of acts as we will see. Nor need there be a prior convention agreed to by both sides that one side violates. There may be such a convention, but it is enough that one side sees a certain kind of behavior as wrong and communicates with the other side about stopping it. The retaliator communicates by both issuing a threat and offering the enemy a sample of the threatened behavior just in case there is doubt about the threatener's intentions.

It should be obvious that the aggrieved side pays a high moral price for engaging in across-the-board escalated retaliation. Particularly if the retaliation is a first-resort response, questions arise about the retaliator's motives. Is the side suffering the abuses so shocked that it desires immediately to express its shock in the strongest possible retaliatory way? Or is that side responding quickly because it is anxious to play the same dirty game, since it sees some gain in store for it? Paradoxically the same shock at the enemy's dirty tactics that tempts the other side to adopt a retaliatory posture should also express itself as a revulsion at acting as the enemy does. If the enemy's behavior is that appalling, it ought not to be so easy for the victim nation to pitch in and get its own hands dirty.

Because the victim nation should show some reluctance in participating in a dirty war, and because it does not want its motives impugned and, further, because each kind of immoral act the enemy commits needs to be looked at carefully, it should hesitate before retaliating in any way. As we have suggested already, the first step it ought to take is to document the enemy's immoral behavior. For obvious reasons, it is preferable that this documentation be done by some neutral body such as the Red Cross or Amnesty International. After that, it is time critically to assess the enemy's behavior. Aside from documenting the enemy's immoral acts and then urging neutrals to apply political and economic pressure to stop this behavior, it is difficult to imagine going beyond that to retaliation. So let us consider what more might be done short of retaliating. Boycotts of those bystanders who still do some business with the enemy could be started. Or, a nation could try to act militarily with greater vigor. However, neither of these actions nor any like it will satisfy. Very likely boycotts are already in place and the military would already be doing all it could to win the war. Thus the suggestion that something more should be done would probably come down to whether retaliation should be tried.

Yet as unsatisfactory as not retaliating seems, retaliating seems even more so. What is the victim nation to do, kill all of the enemy sick and wounded in its hands? To do less would hardly seem like retaliation or at best be doomed to being ineffective since the enemy is doing more. The enemy, we are assuming, is engaged in a general campaign of killing useless prisoners. So must the other side match its enemy killing for killing, or, if it has more prisoners, at a higher ratio? That seems intuitively monstrous. However, the task here is to come to a decision on the critical level. That being so, a strong argument against killing the sick and wounded in the victim nation's hands is that there is little assurance that the retaliation will work and, thus, that the preferences of the victim nation's personnel in enemy hospitals will be honoured. If it happens that the retaliatory response has no effect on the enemy, what is the victim nation to do? Continue the killing?

At this point the following suggestion is helpful. Whenever it makes sense to do so, retaliations should have time limits placed on them. Some retaliatory acts or activities naturally have limits. If the enemy commits a single immorality, it makes sense to respond with a single act. Yet, as we are considering it, if the enemy is engaged in an ongoing immoral practice, it makes sense to retaliate, but to do so, for example, for a week or a fortnight. The time limit would be made explicit. It would be counterproductive, no doubt, to make it explicit in the mass media, thus forcing the enemy, if he were tempted to back down, to appear to be doing so. Rather, the time limit would be communicated through diplomatic or other private channels in order to be kept as quiet as possible. The enemy would be given every opportunity simply to stop killing the sick and wounded. Such a suggestion then might make retaliation more palatable to some people. Before dealing with the question of what is to happen if the enemy fails to respond to this form of retaliation, it is helpful to consider another suggestion for making retaliation more palatable.

This suggestion cannot be applied easily with our sick and wounded case, but it can with certain prisoners. As we have said, our hypothetical enemy is also killing prisoners, except those it feels it can put to useful work. Let us assume, in our example, that some of the enemy prisoners the victim nation is holding can be identified as ones who have themselves killed prisoners. Let us also assume that a still larger group of the enemy prisoners can be identified as belonging to the notorious SS units in their military. SS military personnel, it is known, carry out immoral orders of the kind we are considering with greater vigor than the rest of the enemy's military organizations.

One common argument against retaliatory policies is that they burden the innocent rather than the guilty. Here the analogy is to those classified as innocent in civilian criminal trials. In such trials, the accused is said to be innocent of a murder even if it can be proved that he has committed another murder. He can be convicted only for the crime he is charged with committing, not for another one. In this sense, all of the prisoners in the victim nation's hands are innocent of those moral atrocities committed after they have been captured and on account of which the victim nation is considering adopting a retaliatory stance. Nonetheless, since some of those in captivity have in fact killed other prisoners (sick, wounded, etc.), while other SS prisoners simply belong to the group that is heavily implicated in such immoral acts, these prisoners can be said not to be as innocent as the regular army captives are. Thus, if these less innocent individuals are chosen for sacrifice in the planned acts of retaliation, the original charge that innocents are suffering for the crimes of others loses some of its bite.

Having sweetened the product somewhat, has retaliation been made palatable enough morally to permit it to be used to try to stop enemy attacks upon sick, wounded and other prisoners? Certainly, in a desperate moment, if retaliation must be tried it would seem more appropriate to victimize the SS prisoners rather than the other ones. However, our reluctance to kill prisoners is so great that we tend to look for other options. In this connection Walzer suggests that during World War II the French underground could have feigned killing 80 Germans in response to the killing by Germans of 80 French underground prisoners.[6] Such a trick might work once, if the war is nearly finished and if there is much confusion within enemy ranks. But during a long and protracted war, once the true story were told or was uncovered, one's credibility as a retaliator would very likely come into question. Walzer's suggestion, then, does not seem to help very much in deciding about retaliatory policies.

At this point another possibility needs to be considered. In theory there is no need when retaliating to match each atrocity with a retaliatory act. What is being responded to and the response can be different but linked, nonetheless, through the communicative act that makes a retaliation what it is. If a nation wants to, it can say that some historical site was bombed as a response to the killing of its sick and wounded in the enemy camps. That might not be the most appropriate response but some dirty act needs to be identified as a substitute for the more abhorrent one of killing sick and wounded. The problem here, unfortunately, is that as clear as it might be made that this act is retaliation for another quite different one, the enemy is likely to be provoked into retaliating in response to the original retaliation. Thus if a historical site were hit, the enemy could hit a historical site in response. Not only that, in retaliating against the retaliation, those who were to get relief by means of the original retaliatory act would continue dying. After a while, the retaliatory activity would take on a life of its own as one layer of retaliatory activity came to be piled onto another. It seems difficult, therefore, to justify any sort of retaliation in dealing with an enemy who is killing sick and wounded. Whichever form of retaliation is envisioned, the results seem unsatisfactory. Even the expedient of retaliating by killing enemy sick and wounded over a brief period of time seems both overly brutal and ineffective. Why, it might be asked, would anyone expect an enemy to respond to a temporary brutality on the part of its opponent when he has already adopted such brutality as an ongoing policy? Rather than respond in a positive manner, that kind of enemy would simply expect its opponent to do the same sooner or later.

So once again retaliation appears to be a flawed policy in these kinds of cases. In the end, it is flawed for another reason. Although by not retaliating a nation risks alienating some of its own people who demand that more should be done, there is an advantage in maintaining the moral high ground. The retaliatory posture, even if safeguards such as putting time limits upon retaliation are employed, is at least somewhat a compromised one. Refusing even to retaliate, in contrast, makes it clear to everyone how the one side differs from the other. Now this point is not about posturing in the sense that some people and groups pretend to be acting morally when they are not. By actually not retaliating, a nation shows its people what its standards of morality are. In these contexts, holding the high ground not only keeps the victim nation from perpetrating moral wrongs, it also helps it keep in view its sense of purpose and, in that way, possibly, helps it to fight more effectively.

Up until now, the assumption in discussing responses to a grossly immoral enemy have been bracketed between doing nothing and escalated retaliation. There is, however, a response beyond even escalated retaliation: viz, turning the retaliated act into regular military policy. This is what happened in World War II when the British at first retaliated for the raids on London by bombing Berlin and then later simply developed a bombing habit. A raid on Hamburg or Berlin late in 1944 or early 1945 could hardly be retaliatory in nature. By then, bombing enemy cities was just a way of doing business. Anyone involved in retaliating faces the real problem of what to do when the retaliation has failed to achieve its announced results of stopping some perceived immorality by the other side. A rough and ready test of whether one side should retaliate in the first place is to ask: Are we willing to continue these kinds of acts to the point that they become policy should the retaliations fail? Our reluctance even to think seriously about killing prisoners, the sick and the wounded on a regular basis, just as the enemy is doing it, is an indication that a policy of retaliation should not have started. Perhaps the only kinds of retaliatory acts that might pass this test of permanent policy are those in response to enemy atrocities that give him a clear military advantage. For example, if our pilots are shot at while parachuting to safety, retaliation might be in order. Tic-for-tac shooting at their parachuting pilots might keep the enemy from gaining the advantage here even if it does not stop him from continuing to shoot at our people.

Thus, with perhaps a few exceptions, retaliation makes for bad policy. It probably does not work very often, and against an enemy who is acting immorally on a massive scale, will almost never work. Instead, it is best, in responding to most atrocities, to adopt policies on the command level that dramatize the difference in moral behavior between the two sides. This we have said is important because of how allies and neutrals observe the situation. It is also important because of how one's own people will view themselves. They will more readily comprehend whatever moral rules they are asked to follow if what the other side is doing is clearly different from what they are asked to do. No doubt, between the command level and implementation in the field, there will be a certain amount of slippage. Military personnel who are angry at what the enemy is doing may be strongly tempted to practice local and unauthorized retaliation. Some tolerance here will probably be in order. A soldier in battle who stabs a wounded enemy soldier, who previous to being wounded had stabbed the first soldier's wounded best buddy, could be excused for what he did. However, if too much slippage is allowed, a military organization could quickly become as brutal as the enemy without even going through the stage of offering the enemy a chance to respond to a formal retaliatory proposal. Morally that would be unfortunate.

CHAPTER 8
Weapons of war

1 Forbidden weapons

The list of weapons forbidden by the United States Army's *The Law of Land Warfare* (FM 27–10) is distressingly small and in places anachronistic. Under the heading number 34 we are told about some of these weapons:

> Usage has, however, established the illegality of the use of lances with barbed heads, irregular-shaped bullets, and projectiles filled with glass, use of any substance on bullets that would tend unnecessarily to inflame a wound inflicted by them, and the scoring of the surface or the filing off of the ends of the hard cases of bullets.[1]

A little later we are told under heading 37 that it is especially forbidden to employ poisons or poisoned weapons. Nonetheless, *The Law of Land Warfare* adds:

> The foregoing rule does not prohibit measures being taken to dry up springs, to divert rivers and aqueducts from their courses, or to destroy, through chemical or bacterial agents harmless to man, crops intended solely for consumption by the armed forces (if that fact can be determined).[2]

Under heading 38 it also adds that legally the US is not party to any treaty prohibiting or restricting the use of poisoned weapons.[3]

Other documents either add items to the list of forbidden weapons or put some restrictions on acceptable ones. Mines, for example, when laid at sea or on a potential battlefield, need to have their locations marked so as to be recoverable. Either that or they must somehow deactivate themselves after a period of time.[4]

Small as the list of forbidden weapons is, there is a suspicion that it would become smaller if 'the other side' used some of these weapons first. However, retaliation is not the only factor that feeds this suspicion. Technology apparently drives nations first to conceive of, then make and finally use weapons that, when first conceived, seemed too bad to be true. Once in hand, these new weapons, (e.g., fuel-air bombs, explosives filled with flechettes – i.e., metal darts) make the old ones seem less immoral than before. In the end, the further suspicion surfaces that the list of nasty weapons recognized by nations as immoral roughly corresponds to a list of nasty but relatively useless weapons.

These discomforting thoughts tempt the response: 'Yes, but the weapons in fact forbidden by military forces are not the same as those that ought to be; and any study of military ethics ought not to get overly engrossed in its concern for the former when its business is to deal with the latter.' The objection is well taken. A longer list of immoral weapons could be prepared. It would surely include most area weapons such as conventional and nuclear bombs and shells as well as poison gas and germs. Another class of weapons for our expanded list would be certain target weapons such as high-powered rifles of small caliber whose bullets tumble once they enter the body of their victims. Understandably, damage done by these bullets is often greater than that done by larger bullets that go cleanly through the body. Other nasty weapons such as napalm could be added to the list. In addition, the places where weapons of any kind are forbidden could be extended. Not only the sea beds, the Antarctic and outer space, but cities, towns, and villages could become no-fire zones. Farmers, of course, would argue for having their crop lands added to the no-fire zone list since they would not want land wars fought in their corn, wheat and rice fields. The weapons and places where these weapons could be employed morally would then be such that land warfare, at least, would only be fought in the desert and other isolated locales. Air warfare, dog fights and such, could rage almost indiscriminately as long as the air marshals made certain that the losers in these fights did not deposit too much debris on the city streets. Sea warfare would be even more unrestricted. Admirals could enjoy their wars with no fear of moral reprisal except, perhaps, from fishermen.

Undoubtedly, restricting weapons use could lead to a far more humane war than the kind fought by civilized nations. Presumably, even more restrictions could be imposed. Wars could be completely ritualized. They could be fought by gladiators from the participating nations. The winner could be declared to be the side with the most gold medals in such contests as simulated aerial dog fights, target shooting and helicopter maneuvers.

So it is rather easy to make a distinction between those weapons military forces in fact forbid and the ones they ought to. But this distinction can be made unrealistically or realistically; or, to put it differently, it can be made as we have just done, in a way that will not be taken seriously or in a way that might be.

At this point, in order to avoid confusion, it is useful to distinguish between the noun or name 'realism' and the adjective 'realistic'. Realism, as we discussed it earlier, is a theory about the absence or silence of ethics (and/or law) during war. It is a theory to be contrasted to just war theory since it claims that war falls outside the ethical arena. In contrast, as we are using the adjective 'realistic', it has a sense within, not outside of, just war theory; and within that theory it is to be contrasted to 'unrealistic' or 'idealistic'. Actually there are two senses here, both of which are relevant to some other part of this study but only one of which is directly relevant to the issue of weapons use.

The sense of 'realistic' that is only indirectly relevant has to do with the circumstances of war. This is the sense that prompted us to appeal to tolerance in the previous chapter. War, no matter what weapons are used, is a messy business where frightened, irritated and tired people on the one side try to do terrible things to those on the other side who are equally frightened, irritated and tired. In such situations, it is a mistake to invoke too many moral rules, and to try to put too fine an edge on them – thereby making them unrealistic.

The other sense of 'realistic' is concerned with issues. It speaks to the seriousness of the issues that precipitate and sustain wars. Right or wrong, nations do not normally go to war over issues that they perceive to be minor. Whether these issues are political, economic, religious or whatever, it belittles their perceived importance to suppose that nations would fight ritualistic-like wars to settle the disputes at hand. One's religion is just not that important to a person if he loses it because he accedes to using only a few of the weapons at his disposal to defend it. If it is important to him, he will do all he can to support it and expect his enemy, if he is defending his religion, to do the same. If he is willing to die for his religion, it will not make much difference to him what killed him and what he used to kill his enemy.

If people are deadly serious about certain issues, as they surely are, and if there is no trans-national authority strong enough to control weapons use, as there surely is not, then there will hardly be any limitations on the kinds of and the places where weapons will be used in war. At best, control of weapons will be marginal. But marginal control is better than none at all; and, with some effort, control may be extended marginally beyond the margin. So even though no one should be sanguine about weapons control given the seriousness of the issues that divide humans, it is still worthwhile to look at some weapons systems so as to get a sense of which ones, realistically, can and should be controlled.

II Mines and other cruel conventional weapons

There are a variety of ways to classify weapons systems. This is the way John Keegan does it.

> Crudely, but I think meaningfully, one may distinguish three sorts of battlefield weapons: the hand weapon – sword or lance; the single-missile weapon – musket or rifle; the multiple weapon – machine gun or projector of toxic-gas particles.[5]

For our purposes, although this classification system is a bit too crude, it can be incorporated into one which will serve us better. This system construes weapons more broadly than Keegan's does and then identifies them in terms of one or more of three functions they perform. The first function is *locating* the enemy. This function is often but not always performed prior to the others. Even when large nuclear weapons are deployed, the enemy has to be located first, at least roughly. So some locating instrument is generally employed successfully before the second function, *reaching* the enemy, comes into play. It is one thing to have the enemy in view and quite another to be in position to deliver a blow that will harm him. Reaching instruments, however, come in two modes: viz., transportation and delivery. The transportation instruments, airplanes, tanks, trucks and ships, bring the enemy close enough so that the delivery instruments, rifles, cannons, flame throwers, can be utilized.[6] It is within this category of delivery instruments that part of Keegan's classification of weapons fit. Some of these instruments will deliver single, while others will deliver multiple (or area), missiles.

Having located and reached the enemy, the *disabling* function of weapons comes into play. Some examples of disabling agents are bullets, shrapnel, heat (flame), radiation, gas, air pressure (from an explosion) and germs. The disabling agent is literally what damages the enemy physically, chemically or psychologically so that he can no longer function as he usually does. When a weapon can perform more than one of these functions, we speak of it as a weapons system. The most basic weapons system is the human body in that it locates the enemy with its eyes and ears, reaches him with its legs and arms and disables him with its hands and feet.

Some thinkers about a future conventional war between technologically developed nations have come to realize that such a war would be a fearsome thing. Weapons developments in recent years have taken place across the board in locating, reaching and disabling the enemy. In many cases, this development has not always just been incremental. More than one writer has called it revolutionary.[7] In effect what has happened is this. Locating instruments have extended battle so that it can take place at night as well as daytime. Infra-red, doppler-radar, light enhancing, thermal-imaging and other active and passive devices have made the enemy visible at night.[8] The soldier who in the past might have expected to get some rest at night, now might be asked to continue fighting around the clock. Locating instruments have wrought another revolution. Helped by computers, lasers, television, (doppler) radar and other instruments, the locating function has evolved so that the exact position of the enemy and his equipment can be determined just as the disabling agent reaches him. Such 'terminal' smartness in weaponry requires precise reach as well. And that too has been forthcoming. Miniaturized computers again have come into play in guiding the weapons to the target in response to the messages received from the locating instruments. As if that were not enough, military reach has been extended to cover more area. If a war were fought in Central Europe it could be that all of Germany would be the battleground at one time. Here is the way that John Keegan puts it.

Today, new methods of surveillance, target determination, and weapon guidance mean that ground-launched missiles – like NATO's Pershing and the Warsaw Pact's Frog – can destroy selected targets that are a long way from the point of troop contact. Bombing from airplanes has also been much improved by sophisticated on-board electronic systems. As a result, commanders on both sides now talk confidently about 'fighting across the whole depth of the theater', a phrase that means that targets as far away as the Rhine and as close as the Inner German Border, which separates the two Germanys, might be brought and kept under attack simultaneously. The potential zone of devastation has therefore increased several times over since the trench warfare of 1914–1918.[9]

Conventional disabling agents have kept pace with developments in the areas of location and reach, both qualitatively and quantitatively. Qualitatively, developments in metallurgy and explosives design help put the tank more in jeopardy than it has been in since its appearance on the battlefield. Tanks can be set on fire when hit with projectiles containing 'self-forging' fragments made from depleted uranium.[10] They can also be destroyed by a variety of smart weapons that are 'out-of-sight.'

In one sense all this magnification of destructive power in conventional warfare poses no moral problem, tragic though the death of soldiers in battle may be. The venting of this new destructive power by shooting one 'smart' rocket at a tank, for instance, is basically no more immoral than firing scores of shells at it as might have been done in World War II. Indeed, there is a frightening sense of economy in destroying a tank with one shot. Not only that, in many cases anyhow, smart weapons should be clever enough to go after military targets only. Free-falling bombs aimed at a bridge, to take another example, might not just destroy it, but also part of the village located next to it. Presumably the smart bomb would leave the village alone. In the same vein, modern surveillance techniques ought to be able to differentiate military from non-military installations more readily than in the past and, again, leave civilians alone.

Up to this point we have studiously avoided bringing civilians onto the scene of battle. By doing so, certain moral issues concerned with soldier against soldier and with command decisions were discussed with a minimum of complications. However, the present discussion of conventional weapons no longer affords us this luxury. Although, as has just been observed, some modern weapons make for a cleaner battlefield by not creating a lot of litter, others do just the opposite. Mines have zero reach. Seemingly stupid, they wait for the enemy to come to them. However, since modern mines can be made of plastic materials and therefore are both light in weight and cheap to manufacture, they can compensate for their stupidity by waiting in large numbers over large areas so as to make avoiding them difficult for military personnel, tanks and other military vehicles. Many of these mines are actually not so stupid as they seem since they can be programmed to detonate only after a certain number of vehicles have passed or only when certain kinds of vehicles are passing over them. Smart or not, if military vehicles are troubled by these mines, civilian vehicles, and civilians themselves, will be troubled even more so. If these mines and countless similar bomblets are deployed by airplanes narrowly near a traditional military front line, the harm done to civilians could be kept under control. An army fighting defensively might very well deploy these mines in this constricted manner. By doing so, it could interdict enemy armor and other enemy motorized vehicles quite effectively. However, even a defensive army would be tempted to strew these handy weapons well behind the main line in order to slow down the enemy's reinforcements as well as generally to disorganize him. So even a defensive army would have reason to reach beyond any fighting line with these and other weapons to wreak as much havoc as it could against the enemy. By doing so, the point is, it would also be wreaking havoc against the civilian population since it would be contaminating large livable spaces.

The offensive army would have even more incentive to extend its reach to the limits of its various weapons systems. An army fighting defensively would have more of a problem with refugees than one fighting offensively. Its people would want to flee both the battle zone and the enemy. Roads well beyond the general area where the armor and infantry are fighting would be absolutely clogged with people on foot and in hordes of cars and trucks going one way, and reinforcements, military supplies and equipment going the other way. Even without being attacked in these rear areas, the defensive army's problems would be enormous. However, if the roads, paths, and fields across which all these people and machinery move are mined, bombed, strafed, and, further, attacked by helicopters landing enemy troops in their midst, near total paralysis of movement would result. Military and civilian casualties would be enormous as the offensive army attempted to squeeze the defensive army in an area between the front edge of the battle area and the barrier of near total confusion in the rear.

Having just introduced civilians into the battle area we are not yet in position to say anything very definitive about the morality of this

European Central Front scenario. *Prima facie*, however, the offensive army seems subject to criticism for not only taking advantage of the plight of helpless civilian refugees, but also deliberately worsening it. The defensive army cannot avoid criticism either. Quite obviously, in our scenario, it is using the same weapons as the offensive army. Less obviously the defensive army is subject to criticism by not having prepared for the kind of attack that came. By having done nothing to prepare for war, it invited the attack. It would not do for the defensive army to posture in the direction of the enemy by saying 'You wouldn't do anything like that, would you?' and, thereby, try to hide its military weakness behind a mantle of morality.

Dealing critically with the morality of this scenario is difficult for the following two additional reasons. First, scatterable minelets and bomblets are not easily subject to arms control measures. Developing such weapons, producing and then deploying them by the millions could be done in great secrecy. It would do no good and perhaps great harm for both sides in a potential conflict to outlaw these weapons. The great harm would, of course, come to the side that honored the agreement. All that nation would need to do to lose the next war would be to honor two or three more like it. The no good would come because neither side would likely honor such an agreement. Even here the expression 'no good' is misleading because of the presence of long-term harm. By making the kind of moral agreement that both sides are prepared to violate under the pressure of war, morality tends to get devalued since unenforceable agreements encourage nations to ignore the enforceable ones. The fact that these weapons are not subject to arms control measures even up to the time they have been deployed is no overriding reason for not trying to control them. Production and deployment of certain chemical and bacteriological weapons can also be carried out with a high degree of secrecy, and yet efforts to control these weapons continue. We will have more detailed things to say about these weapons shortly. For now, however, notice two important respects in which these outlawed weapons differ from the new plastic minelets and bomblets. For one, these latter weapons are outgrowths of already accepted military technology. They are natural refinements of instruments that have been killing military personnel for generations. It is difficult, therefore, to outlaw the new, lighter and more powerful mines and bomblets since the line between the old and the new is not so easy to draw. For another, the new weapons have legitimate military uses.

If they can be sprayed in front and on top of a mobile enemy military column, is one to say that they are to be labelled immoral because they do their intended work more effectively than older weapons? The second additional reason the Central Front scenarios cause difficulty has to do with the rich mix of the military among the civilians in the heavily clogged rear area. Indeed, in a gesture of humanity, the offensive army could avoid direct attacks on columns of people and machines moving away from battle when these civilian columns are not intertwined with the military columns going the other way. Still, on many occasions, the columns would not be sortable easily. Whatever the case, the minelet laying operations would likely do as much harm to those moving toward as to those moving away from the main battle areas. Although civilian casualties would be high, so would military casualties. It would therefore be difficult to condemn the offensive army outright since it is, in fact, doing significant military work by laying the mines.

Before summing up and drawing some tentative conclusions about the morality of the new minelets and bomblets, the 'residue' problem needs to be looked at. Sowing mines by hand or even from a ship made it at least possible to locate and pick up the mines after the battle. Those who did the sowing in those ways had more than a rough idea where they put their mines. Besides, since mines a generation ago were made of metal, even if they were mislaid, there was a good chance that they could be located. Further, since they were heavy and expensive, only so many, often more than enough to be sure, could be laid in any one area. Now, aside from the great destruction they can wreak during war because they can be deployed in such great numbers, the plastic nature of the new mines makes them difficult to find after the battle is over. To some extent this same point applies to the millions of bomblets that could be dropped during a war. Some of them will not explode so they too become mines for all practical purposes. Given the nature (and number) of these weapons, the after-the-battle costs of clean-up will be abnormally high. Unless these weapons are made to self-destruct within a certain period of time, lives and money will be lost in clean-up and much time will pass before the mined areas can be restored to peaceful uses.

In summary, the seeming unexotic nature of minelets and bomblets belie the moral problems they pose for us. As we have seen, they perform their disabling function more because of the numbers than their explosive power, although that should not be minimized either. That power can in fact be tailored to deal either with personnel or armored vehicles, so these weapons are not as unsophisticated as they may seem. These weapons perform their functions well in part because they are easily transported to the enemy and delivered to the area where he is located. However, because the area where they can be spread is so great, it is inevitable that they will victimize military personnel and civilians alike. Nonetheless, in spite of their almost random destructive power, we tentatively conclude that it makes little sense to condemn the production, deployment and, in many cases, even the use of these weapons. It is how these weapons can be employed, and not the weapons themselves, that should be subject to censure. It seems that condemnation makes sense (1) when they are used directly over areas where civilian traffic dominates and (2) when no provisions are made for cleaning up mined areas after the battle is over. This does not seem like very much of a condemnation, but realistically not much more can be expected.

Actually one other thing might be expected. If, as John Keegan and others have suggested, conventional weapons systems are far more dangerous today than they were even in the last world war, it is important for the rulers of nations and the people they rule to bear this fact in mind. By coming to realize how horrible a conventional war is likely to be, some nations might hesitate a little more before launching into one.

III Poison gas and biological weapons

People's expectations about controlling chemical weapons such as poison gas are somewhat higher than with some of the weapons we have just discussed. To be sure, poison gas is an area weapon like cluster bombs and mines. Also like these weapons, they come in various forms, some of them being *prima facie* morally more acceptable than others. So called tear gases (e.g., chloracetophenone) are generally classed as non-lethal gases not because they never kill anyone, but because fatalities among those exposed to them are relatively low (2% or less).[11] Other gases fall into this non-lethal category including CS and 'pepper gas', and 'BZ'. Although these more acceptable gases can be confused with the more lethal varieties (e.g., blister types such as mustard gas, and choking gases such as phosgene and, perhaps the worst, nerve gases such as tabun, sarin and soman), they are different enough in their chemical composition and their effects on their victims so that they can be distinguished from one another when used on the battlefield. Further, the lethal gases have a tradition of non-use, established primarily during World War II, to give people some hope that governments will desist from using them in future wars.

It is more difficult than might be supposed to say why poison gas seems to be such an immoral weapon to so many people. Of the non-nuclear weapons, it evokes the greatest sense of horror at what it does and a corresponding sense of disgust at those who use such a weapon. Yet it is certainly not the weapon that has done the greatest harm. When poison gas was used extensively during World War I, it caused fewer than 100,000 deaths and a little over 1,200,000 other casualties.[12] Compared to the almost 9,000,000 men who were killed in that war and the over 21,000,000 who were wounded, it is obvious that the bullet, the artillery shell and other disabling agents did more harm. Nor is it even obvious that it is such a potent military weapon, the kind that would threaten mankind, that it needs to be condemned on that account. In fact, even though it was used extensively in World War I, it did not by itself or in conjunction with other weapons, get either side out of the quagmire of the trench warfare on the Western Front. With greater reach, because of the use of modern aircraft and rockets, and with more powerful forms of poison gas, this weapon is surely more powerful today than it was during World War I. Yet, the same can be said of many other weapons. Finally, nor is it obvious that it is impossible to defend oneself against gas attacks. Such a defense is not easy to arrange, but both military personnel and machines can be prepared to keep the effects of a poison gas attack at a minimum or at least to manageable proportions.[13]

So why the horror? In part it may be because the effect of poison gas can be so vividly imagined in two senses. Although poison gas may kill quickly, it is not perceived to be so merciful as a bullet that can kill a person before he knows what has hit him. The victim sees himself as suffering needlessly before dying. Also, the suffering itself – the inability to breathe, choking, vomiting, convulsions – is easily imagined as a traumatic way to die. Other more conventional weapons may also not kill immediately but, often, they knock one senseless so that when death comes a person is unaware of its arrival. Poison gas may or may not, on average, cause more pain and suffering than do the more familiar metal disabling agents, but it certainly gives the appearance of doing so. So much so that it is safe to say that most military personnel would probably prefer, if they must choose, to be killed or disabled by some other means.

The other part of why poison gas holds people in horror has to do with the kind of area weapon it is. The artillery shell is an area weapon by virtue of the way it is delivered to the enemy. The artillery piece aims at a target but, unless the shell is exceedingly smart and the enemy has been located precisely, the target for all practical purposes is the area somewhere around the enemy. The shell fragments are area weapons in a different sense since once the shell explodes it sends them out from the point of explosion in all directions.

Poison gas is cruder than a shell or its fragments, although not necessarily because of considerations having to do with delivery. It can come to the enemy just as the metal disabling agents do via a shell, or a rocket or bomb. Its crudity comes from its ability to spread beyond the area that a shell or bomb hits. It spreads to where the wind or gravity carries it to do damage over wide areas. Its (area) coverage is not self-generating the way germ weapons are. These are the crudest area weapons of all. With germ weapons the multiplying disabling agents can spread in such unpredictable ways that these weapons have limited military value. Another consideration that limits the value of biological weapons, especially in a short war, is that germs, viruses, etc. need a considerable amount of time to do their dirty work. Poison gas, in contrast, is not only quick acting, but as an area weapon is just barely controllable. With modern delivery systems, it can be aimed accurately at military targets, but with its capricious tendency to spread itself around, it would almost unavoidably harm civilians. This is especially true in a modern war. In World War I, gas was primarily employed close to a static battle line peopled almost exclusively by soldiers. Since, as we have seen, the effective reach of modern armies and air forces is so great that it would be difficult even to identify a battle line, civilians would inevitably be gassed. In fact because it is so crude, and because civilians would be less able to defend themselves against a poison gas attack than military personnel (since the latter have more training and equipment to protect themselves), civilian casualties would likely be extremely high if it were used.

Thus far we have been concerned with the use of poison gas only within the context of both a high-intensity and high-technology war.

And the conclusion to which we are gradually moving is that the use of this weapon is morally unjustified. Our conclusion is tentative since, as yet, we have not investigated as fully as we will in the next chapter the status of civilians of war. As of the moment, we are assuming rather than arguing for the thesis that at least some civilians are not worthy targets of military action. It is that assumption (or intuition) and the earlier argument that poison gas is easily visualized as a horrible weapon that has allowed us to put this weapon on the taboo list.

More than likely it would remain on that list in a medium or low-technology war, that is, one taking place between client states of the super-powers who are given less than the state-of-the-art weaponry. Even if the weapons used in such a war were not all they could be, both in quality and quantity, the depth of the battlefield would likely still be large and fluid enough to cause many civilian casualties. It would, of course, make no difference what the technological level of the war was to the soldiers. If being gassed is an especially horrible experience, it will be so both for the super-power soldier and for the soldier from the client state. So the moral condemnation of gas should not be affected by the technological level on which the war is being fought.

It does not follow that because a weapon is immoral and thus ought not to be used in the normal course of war that, first, it ought not to be stocked and, second, it ought not to be used in retaliation when the other side has used it first. With respect to stocking immoral weapons, it is tempting to take the high road. If poison gas is seen as an immoral weapon, it seems best not to dirty one's hands by even making the stuff.

The high road might well have been the best to take, if this weapon were easily subject to monitoring. But, as we have seen, poison gas can be delivered by a number of weapons systems. Further, the gas itself can be produced (or purchased) secretly and thus is a prime candidate for status as a surprise weapon. More than that, it is an effective weapon. There are a variety of military reasons an enemy might be tempted to use it. In a high-intensity high-tech war, poison gas could be dropped well behind the forward edge of battle area among the military and civilians we have talked about already. It could also be used in the same kind of war if it were clear that the victim nation's NBC (nuclear-biological-chemical) defense equipment was inferior. Presumably part of the strategy in using gas in such a war would involve the user employing effective NBC equipment so as to move quickly into the area where poison gas has been spread. Yet even if the victim nation were reasonably prepared to defend itself against poison gas attacks and therefore suffered only acceptable losses, it would still be at a disadvantage militarily. Clothing worn to protect military personnel against modern poison-gas weapons is not comfortable. One either wears it for fear of an attack that will come at any moment and, because of the discomfort, works at a lower level of efficiency; or, one takes the risk of becoming a casualty by not wearing it. Thus, if a nation had chosen for moral reasons not even to stock poison gas, it would be depending upon the enemy to practice restraint in the use of these weapons even though it knew that the enemy knew it could gain clear military advantage by not restraining itself. The non-stocking nation would be relying upon its enemy's moral integrity to keep it from succumbing to temptation.

If the enemy's integrity were somewhat suspect, it might be thought that world public opinion could help improve it a bit. Possibly, public opinion could help in certain military settings. In a protracted war, such as the one between Iraq and Iran, the (poison gas) user nation might back down under such pressure. Were public opinion to fail, the world's more powerful nations could, additionally, apply economic and political pressures to stop the offending nation's immoral practices. However, in a blitzkrieg war fought in central Europe there would neither be more powerful nations to apply any pressure, nor would there be time for such pressure to build up to do any good. There would only be the fact that one nation won the war because it used poison gas in conjunction with other weapons. The high road of not even stocking such weapons thus seems, if anything, to encourage nations to use poison gas and other chemical weapons. Paradoxically, in order to keep nations from not using such weapons, it may be necessary to keep them in stock.

This conclusion will make many people uncomfortable. They will argue that weapons deployed will be weapons used. There may be exceptions to this 'rule' but, they will insist, most nations will succumb to temptations to use these weapons if they can get their hands on them. Overall, they will then argue, it is best to eliminate them from our arsenals completely.

As a wish, such a conclusion makes much sense. As a realistic moral prescription it fails. It is true that the kind of deterrence position we are advocating runs certain risks since there is something right about the 'If you've got it, you'll use it' slogan. If the major military powers develop and deploy such weapons to deter one another, the technology and materials, and perhaps even the weapons themselves, will get into the hands of lesser powers. So in addition to being tempted to use the weapons themselves, there is also the danger that one or two of the supplier's client states will find an excuse to use them. A desperate, small nation might, for instance, feel justified in using poison gas in an attempt to thwart what Walzer calls a 'supreme emergency.'[14] But even if it were granted that a deterrence policy runs these risks, the argument here is that an abolition policy runs the greater risk of encouraging violations in a war that could very well be one sided. Thus the poison gas abolitionist simply cannot say that he recommends to everyone the wonderful world where no one can use these weapons because no one has them. He must realize that the realistic choice is between unilateral abolition, with its dangers, and deterrence with its. We are arguing that the dangers of the latter opinion are fewer and less serious than the former.

Our abolitionist could respond by saying that his fears about the use of gas are not restricted to what other nations do. He may not trust those in his own military establishment, thinking them to be either mavericks or simply irresponsible. What is likely to happen, he thinks, is that in spite of the deterrence stance of the major nations, somebody, on one side, possibly one of his own leaders, will do something stupid and war with poison gas will result. That result, he might add, needs to be avoided since a war between major powers all of whom are amply stocked with poison gas would represent something close to a worst possible scenario.

No doubt a bilateral poison-gas war between major military powers would be horrible. Yet, there is little to be done about the maverick or irrational behavior of those making weapons decisions in a potential enemy nation. About all one can do is avoid stocking these weapons in amounts beyond what effective deterrence demands. More can be done about controlling one's own military leaders and institutions. Indeed, it has been the burden of much of what has been said in Chapters 3 and 4 that quite a bit can be done. In these chapters we said that mistrust of the military establishment can be lessened if this establishment is not kept isolated from the rest of society. In part this means that the military should be encouraged to criticize the society for what it is not doing in the area of defense and be receptive to outside criticism in turn. We can now see another reason for the wisdom of this policy of non-isolationism. Since there are strong arguments for a deterrence policy in dealing with such weapons as poison gas, since such a policy puts dangerous weapons in the hands of military people and since these people might do foolish things with these weapons, it seems wise to adopt a policy of watching them carefully. The solution to weapons-in-hand problems is, as we said earlier, monitoring the military, not emasculating it.

About poison gas there remains only to decide whether, once stocked, these weapons should ever be used. As a first-strike weapon a negative answer is appropriate for all the reasons given above and for an additional one. In these modern times a nation should be able to find more than enough morally acceptable weapons to satisfy its military requirements. About the only occasion that it might seriously consider using poison gas is when it finds itself in a retaliatory posture. Thus far we have not looked kindly upon retaliation as a policy, although we did argue that there are occasions when such a policy might be adopted. This is another such occasion, and for the same reason we gave in Chapter 7. There just is too much military advantage in letting only one side use poison gas. As most texts in this matter recommend, the retaliatory nation should not jump in with enthusiasm with its response-in-kind. Warnings should be issued and demonstration retaliatory acts might be tried. In the end, however, if the enemy did not desist such weapons might very well be used.

If the need to retaliate is premised largely on the military effectiveness of the weapons and if, as it seems, biological weapons are not as effective as poison gas, retaliation with these weapons should either not be permitted or be even slower in coming. Certainly, in a short war there would be no need to retaliate with biological weapons since militarily it would not help to do so. In a longer war, it might depend on what particular biological agents were used but, in any case, a peculiar form of retaliation is available with biological weapons not available with any other. We already noted how difficult it is to control biological weapons. As self-generating instruments of war they can spread capriciously. In fact, one of the dangers of using them is that the germs, bacteria, etc. might eventually spread back to their place of origin. This could happen naturally or their return home could be aided by measures taken by the victim nation. One might even be reluctant to call this form of counter-attack retaliation, since the germs would be direct offspring of the ones sent. It would be a bit like using mirrors to reflect back on the enemy the deadly laser beams he sent in the first place. Call it retaliation or not, it would be a particularly satisfying way to get even with the enemy even if it might not be morally very pure.

IV Nuclear weapons

As we have seen, a modern high-intensity, non-nuclear war may approach the destructive power of a war fought with small tactical nuclear weapons. Using cluster bomblets and minelets, poison gas, fuel-air bombs, a variety of automatic weapons, rockets, artillery, modern tanks and fixed-wing aircraft, helicopters, all sorts of locating instruments and scores of other new and improved versions of older weapons, a startling amount of devastation can be caused in a very short period of time. It is because of this potential for destruction that the discussion in this chapter has until now dwelt on what, if anything, should be done about a few of the more dangerous of the non-nuclear weapons. Yet as terrible as a modern non-nuclear war can be, a nuclear war would clearly be much worse.

A review of the following facts shows why this is so, and also why dealing with the problem is so difficult. First, there are enough nuclear weapons in the hands of the two major powers so that even in a partial nuclear exchange, far more people would be killed in a day or so than all the people killed together in World War I (approximately 30,000,000) which lasted four years and World War II (approximately 45,000,000) which lasted five.[15] Second, not only would the immediate killing, maiming and destruction be greater than in these wars, the aftermath could be incalculably destructive. No one really knows what would happen after an attack of such suddenness, that is, no one knows how disease would spread and how radiation, dust and smoke would affect people and their environment near and far. The destruction could easily exceed the direct destruction wrought by the impacting effect of the weapons themselves.[16]

Third, questions about the production, deployment and use of nuclear weapons cannot be separated from questions about how these weapons are delivered to their targets. It is because nuclear weaponry was mated first to the bomber and then to the rocket that we have a very serious problem at all. Rockets in particular not only have maximum reach but can achieve this reach in minutes. If nuclear weapons were so large that on land they could only be carried by train to the border, they might very well play a stabilizing role on the world scene as purely defensive weapons. In fact, it is just the opposite now. Because these weapons can be delivered any place almost at any time, a potential attacker has every incentive to build as many nuclear-tipped rocket delivery systems for which he can find targets.

Fourth, probably unfortunately, but possibly fortunately, there has been no technological abatement in the area of nuclear weapons/rocket technology. Nor is there any indication that such abatement is about to take place. If anything, technological development seems to be accelerating. Some people had hoped for abatement after the 1972 SALT I agreement and a follow-up agreement between the US and the USSR in 1974.[17] However, development and deployment continued, spurred partly by the same electronics revolution that put increasingly efficient computers, watches, radios and television sets in our homes.

The implication of the fourth fact needs to be appreciated before proceeding any further. The new accuracy inherent in modern intercontinental and intermediate ballistics missiles (ICBMs and IRBMs) makes it, or soon will make it, possible for these missiles to fall within a few feet of their targets. What makes such accuracy possible is not just that science has learned to aim a rocket better when it is launched. It has done that to be sure. But it has also learned, with the aid of computers and other sensing devices, to re-aim the rocket in midcourse and then to make further adjustments of the warhead's fall toward its target with the aid of terminal guidance technology. In a sense, the warhead looks for its target and adjusts its course once it sees where it is. A fixed target like an enemy ICBM or IRBM in its silo is, then, an easy target for one of these smart weapons. Clearly, this kind of smart weaponry can have a destabilizing effect militarily. For the USSR it is especially de-stabilizing since a large percent of its ICBM force is based on land and is therefore vulnerable to a counterforce attack (i.e., one aimed at weapons and military installations rather than cities and civilians – called a countervalue attack).[18] Additionally, since the majority of the USSR's ICBMs are still of the slow-firing liquid-fuel type, this nation has another reason to be nervous about the first-strike capability of the US.[19] It is understandable that the Russians might translate all this nervousness into an attack on the US's first strike missiles in order to preempt a first-strike attack. In turn, the US's nervousness about losing its land-based missiles, certain naval bases where much of its submarine-launched ballistics-missile forces would be located (for repair, refit and resupply), and air bases (where its bomber-launched missile forces would be located) would tempt it to preempt the USSR's preemptive strike.

This kind of scenario is an example of an unstable condition. Such a condition is a period of time, usually measured in blocks of a few years, when at least one potential enemy has a strong incentive to start a war or when conditions are such that a war is likely to start through inadvertence. The scenario portrayed above, caused in part by an abnormal combination of technological advances in offensive capabilities and lack of comparable advances in defensive capabilities, is only one of several possible ugly scenarios. A technological-political unstable condition could be reached, for example, if all nuclear weapons and nuclear production facilities were dismantled, but one side had mobilized overwhelming conventional military forces. More realistically, another unstable condition would be reached if both sides' nuclear weaponry were neutralized (since they could almost simultaneously and utterly destroy the other with these weapons) and yet one side had overwhelming conventional forces at its disposal.

Scenarios featuring still other unstable conditions can easily be imagined. A purely technological unstable condition could be reached if one side could dominate outer space both offensively and defensively. Using outer space to locate enemy ships and military movements, and using it as a launching station for rocket attacks, the dominant nation could attack almost at will. Using outer space to intercept enemy missiles, this nation could destroy the whole of its enemy ICBM and IRBM forces. Such space dominance might even help it to control attacks upon its land mass from low-flying cruise missiles.

A non-technological unstable condition could be reached if one nation placed itself in an economically vulnerable position by, for example, becoming overly dependent upon oil imports. Thus, for both technological and non-technological reasons, nations could find themselves moving toward or away from conditions of instability. If we assume that both sides prefer to get their way by not fighting a war, especially a nuclear one, the moral problem of dealing with nuclear weapons can be put in the following form: What courses of action are most likely to move competing nations away from these conditions of instability that lead them to use these weapons?

As the scenario mentioned above concerned with nuclear weaponry disarmament suggests, not just any disarmament policy will do the job. The best option is not one that moves nations from one unstable condition to another. Nor need we naively assume that just because technology got us all into this nuclear nightmare that the only way out of it is to reject technology or, as a second best option, freeze it in place. These steps might work. Yet, it is also possible for new technology to move nations away from the unstable conditions of preemptive nuclear strikes without necessarily moving to a new unstable condition. One example of how technology has helped already is through satellite surveillance. Knowing what the other side has and what it is doing with its rockets, ships, tanks, etc. tends to minimize technological and tactical surprises and therefore tends to keep some forms of national paranoia at less than critical levels. Granting that new developments in technology may move nations away from conditions of instability, we must also keep in mind that this same technology just might move us in the opposite direction.

It is not possible nor is it appropriate for a work in military ethics to go into great detail concerning the technical aspects of nuclear attacks by ballistics and cruise missiles, and defenses against them. However, it was necessary to say what we have already to show that technological development is a live option in dealing with the threat of nuclear war. Having introduced that option, we are in a better position to appreciate the other options, and then to ask which one is most likely to steer contending nations away from whatever unstable conditions their relationships with one another bring about.

One obvious and easily achieved option is total unilateral disarmament.[20] The virtue of this option is that it automatically spares everyone from the horror of nuclear attack. Under this option, one side, being without weapons, cannot attack, while the other side, not feeling threatened, need not. Of course, the other side could attack in a kind of war of caprice or revenge. But this is hardly likely. Use of nuclear weapons would merely devastate the unarmed nation in a way that would make it unworthy of economic and political exploitation. Presumably the conquering nation would prefer to control its victim while it is intact so as to draw as many raw materials and products from it as possible. To do that, all it would probably need to do is threaten to use all or a part of its arsenal of nuclear weapons. If that did not work, a token attack surely would get the response it wants. Actually this is the cynical sub-scenario associated with the unilateral nuclear disarmament option. The sanguine sub-scenario goes like this. Unilateral nuclear disarmament would not only have the consequence of making nuclear attacks pointless, it would lessen national paranoia. One of the main reasons nations have come to the present (nuclear) unstable condition is fear of the other side's military advantage. Fear leads one side to build its military establishment, and then fear on the other side leads to a build-up on that side. Fear thus builds on fear. Once the cycle is broken by one side disarming unilaterally, the other side, no longer threatened, will no longer feel aggressive. Far from pillaging and exploiting the naked nation it would, more than likely, leave it alone.

Unilaterally freezing technology in place by refusing to update one's nuclear weapons amounts to total unilateral disarmament. It is unilateral disarmament over time since with technology developing as fast as it is, and with wear and tear on existing weapons, these weapons would become useless in a few years.

Although a unilateral technological freeze is not an option significantly different from unilateral nuclear disarmament, bilateral nuclear disarmament is. If this form of disarmament were total it would have to be multilateral, rather than merely bilateral since, of course, many nations have such weapons. More realistically the disarmament could be partial, with the superpowers divesting themselves of only most of their nuclear weapons and missiles, thereby not losing their superpower status. Presumably the merits of this option (and the unilateral one) are that it would ease tension and would take the pressure off the economies of the contending nations by cutting military costs. More resources would, thereby, be made available for the benefit of ordinary people. A watered-down version of this option would simply control rather than stop the development and deployment of nuclear weapons and missiles. One possibility would be to allow a nation to add a new missile to its arsenal if it deactivated two or more older ones. Another possibility would be to allow the deployment of a new missile every five or ten years. A particular example of this bilateral option is worth mentioning since it represents a way of dealing with nuclear/missile threats that was actually chosen by the two superpowers. SALT I is an agreement between the US and the USSR made in 1972 that essentially forbids (with minor exceptions) these two nations from building defenses to protect nations from ICBMs. It also places restrictions on the number and kind of nuclear missiles that each side could deploy. On the US side this agreement confirmed the legitimacy and led to the popularity of the doctrine of Mutual Assured Destruction (MAD).[21] Although this doctrine was in part historically an outgrowth of SALT I it actually represents a separate option. In part it is separate because MAD theorists need not, although they might, believe that bilateral agreements are worth pursuing. In essence, MAD is the view that the two superpowers have reached parity at such high destructive levels that, if each acts rationally, each will be deterred from attacking the other with nuclear weapons. There is a corollary concerned with defenses that goes with this doctrine. It is assumed by MAD proponents that nations should be encouraged not to clothe themselves defensively. It is as if by going naked defensively, even nations led by somewhat crazed leaders will not fail to see the madness of starting a nuclear war. A second corollary is that the other side believes in MAD.[22]

Critics of the MAD doctrine have developed another option for seeking stability. MAD proponents, these critics have argued, may not have fully appreciated the potential of defensive weaponry against the ICBM.[23] Actually the MAD argument is not just that defensive technology against ICBMs should be eschewed but that it is not promising at all since it cannot possibly work against 100% of the enemy's missiles. What good, the MAD proponents argue, is a defensive system if it stops only nine out of ten nuclear bombs from hitting a city? The margin of error in a nuclear war, they argue, is zero. However, defenders of an anti-ballistics missile (ABM) system retort that defensive missiles do not have to be perfect to serve an important stabilizing function.[24] They admit that even the best layered defensive network (i.e., an arrangement where ABM missiles first attack ICBMs soon after they are launched, then attack those surviving the first attack in a second layer of defense roughly at midcourse of the attacking missiles' flights and then, finally, attack the few remaining warheads in a third layer of defense just before they hit the targets) may not be perfect. But they argue that since an ABM system would be used initially to defend only certain geographic points (e.g., where the ICBMs are located) it does not need to be perfect. At least in the initial stages of development all that would be required of an ABM system is that it protect enough of the ICBMs so as to make an attack on them less inviting. Since it is likely that smaller, last-ditch non-nuclear defensive rockets will be cheaper to build than offensive weapons systems, building defensive weapons should serve as a disincentive toward building more ICBMs.[25] Thus if a potential aggressor attempted to overwhelm the enemy's defensive weapons by brute ICBM power, the defender could counter not by building his own more accurate and/or more brutish ICBMs, but simply by adding extra non-nuclear defensive rockets to his arsenal. Even if it turned out that the defensive weapons were not cheaper, these weapons would, the ABM proponents argue, have the moral advantage by denuclearizing weapons deployment to some extent.

The last option in dealing with the threat of nuclear war combines certain features of the defensive option and the MAD option. Like the former, it both envisions spending more money to develop new weapons and places little confidence in bilateral agreements, but like MAD it emphasizes offense rather than defense. Adherents of this option are impressed with technology's advance. They see that today's deterrent rocket is tomorrow's easy target of increasingly accurate counterforce missiles. Even if today's rockets could be launched the moment it appears that a nation is under attack, these (neo-offensive) and other theorists see this as a trigger-sensitive and, therefore, an unstable arrangement.[26] So far the proponents of this option agree with those for defense. However, they disagree by recommending that technology develop less vulnerable offensive weapons. In essence, these weapons must be movable and therefore probably smaller than those made in the past. They could also be more sophisticated especially if the enemy began to take defensive measures against them. Ways of tricking or overwhelming the defense would have to be devised. It is debatable whether these new offensive rockets would have to be more accurate. If they were targeted against cities, accuracy would not be important. If they were assigned counterforce tasks, accuracy would be. More than likely the advocates of this offensive option would want to incorporate accuracy in their new missiles since, even as a second-strike weapon, fixed military targets such as radar installations, air fields, naval bases and unfired enemy ICBMs could always be found and hit. For all practical purposes, this option can be thought of as a variation of the MAD theory since most of its advocates believe that the new less-vulnerable mobile missiles will allow deterrence to work in spite of rapid changes in technology.

V Avoiding instability

Having laid out the options, we are now in a better position to answer the question posed earlier: viz., what courses of action are most likely to move the contending nations away from those conditions of instability that lead them to use nuclear weapons? This question clearly presupposes that fighting a nuclear war to win is not an option. It may not be out of the question in the distant future in a Buck Rogers scenario fought in deep space. However, given present and near-future technology and given the roughly tens of thousands of nuclear weapons in existence, and given also the generally acknowledged difficulties of keeping even a limited nuclear war from continuing and escalating and, finally, given the great damage these weapons can do, we think our presupposition is well founded. This is not to say that no person or nation disagrees with this presupposition. There may be powerful people within the nuclear-power nations who hold to a doctrine that a nuclear war can be won through the use of more accurate missiles and cleaner bombs.[27] Indeed, someone with this frame of mind could look seriously at all of the same options we are considering and ask himself which one is most likely to give his side the winning edge. That this is possible makes it clear, if it is not already, that these options are not means to only one end: viz., the stability of peace. This is just another way of saying that no option has some moral certificate of approval as the one most likely to move nations away from conditions of instability. So long as even an advocate of the winning edge can also be an advocate of a bilateral disarmament agreement (since he figures that the other side will stupidly comply with the agreement more readily than his side), so long will each option have to be looked at with suspicion.[28] Suspicion is also in order concerning any recommendation if we keep in mind the fact of rapid technological change. Given this change, today's correct recommendation might turn out to be worse than worthless tomorrow.

With these suspicions in mind, it is perhaps best to proceed in a piecemeal fashion. Instead of looking for sweeping and grand answers to our 'What courses of action?' question, it might be best, first, to identify the specific causes of instability we are trying to avoid and, then, second, to identify specific courses of action that show promise of getting the job done. There are, in fact, many specific causes here. In this chapter we are concerned only with those causes as they directly relate to nuclear weapons and their delivery systems. Other causes of the unstable conditions that nations find themselves in at present include mistrust based upon historical events, basic ideological differences, lack of communication between nations and self-serving military industrial complexes. Unfortunately most of these causes cannot be made to disappear overnight or even in the foreseeable future. But specific to the weapons themselves one candidate as a cause of instability is the sheer number of nuclear bombs and missiles in the arsenals of the two superpowers.

It might seem obvious that the appropriate recommendation would be to get the nations guilty of this overproduction to deactivate many of their weapons either through some bilateral agreement or through some unilateral action. In fact some unilateral action along these lines has taken place on the western side of the nuclear face-off.[29] The US at the moment has 30% fewer nuclear warheads in its arsenal than it did 20 or so years ago. Many bombs fitted to older rockets and many others configured as mines close to the front lines have been deactivated. The fear concerning these latter weapons is that if a war on the central front were to start, and if, as is likely, the eastern bloc armies would move west quickly, the western powers would be forced to use these weapons or lose them. Clearly, therefore, the west's unilateral disarmament steps move the superpowers away from instability by at least placing certain nuclear weapons at arms' distance from the front edge of the battle area.

In truth, this commendable change in the deployment of nuclear weapons is not as much of a gain as it seems. To some extent nuclear weapons that would have been delivered on top of the enemy by close-in artillery or underneath him by mines are now replaced with ones delivered there by missiles or other long-reach weapons. Even with the old weapons gone, there are still plenty out there. Understandably this fact prompts those in favor of further unilateral nuclear disarmament to remind us all of the irrationality of having the power to kill the enemy for five, six, or more times over. No doubt there is much irrationality to be found in the nuclear-war arena. But the degree of irrationality present seems far greater than it actually is if we focus attention exclusively on the weapons themselves, rather than on the targets they are intended to destroy. In the recent past, we have been experiencing not only a proliferation of weapons but of targets. Targets proliferated as the military's reach at first extended out crudely so that large and distant chunks of real estate such as cities were turned into targets, and then later extended in a more refined way to grant small and distant chunks of real estate such as missile silos the same status. Also added to the list of targets were air and naval bases, radar installations, communication centers, control centers, dams, power stations, you name it.

It is much easier to get rid of ICBMs than the targets so it seems rather pointless to focus attention away from the rockets to something else. Still, if there were only a hundred targets the rockets could aim at, there could be no need for thousands of missiles. Targets create the need for missiles, and until the number of things that can be targeted can be made fewer, it is going to be difficult to convince the major nuclear powers to cut down on the number of missiles they deploy. So how can the target numbers be lessened or at least be made less worthy targets than they are now?

The simplest way is to dismantle such targets as land-based missiles and military installations. But that simple way represents nothing more or less than unilateral disarmament, a way which both (all) nations are reluctant to walk down. As we said earlier in this chapter when discussing conventional weapons, it is not realistic to suppose that nations will lay down all or most of their arms unilaterally (or bilaterally) in the foreseeable future, given deep ideological differences, mistrust based on historical events, etc.

A less simple way, in fact a very complicated and expensive way, is to phase all fixed land-based missiles out of military arsenals and replace them with the peripatetic missiles mentioned above under the neo-offensive weapons option. This proposal would represent no movement away from the present unstable condition insofar as total missiles are concerned if, for example, a thousand fixed-in-place and potentially targetable US missiles were replaced with peripatetic ones carrying an equal number of warheads. Presumably, however, if the mobile missiles on both sides could not target each other, there would be less military utility in replacing some of the old missiles with new ones on a one-to-one basis. So long as there were more old missiles deactivated than new ones deployed, some movement away from instability might be achieved. There would be movement here both because there would be fewer targets to shoot at and because the temptation to start shooting at elusive enemy missiles would be lessened.

A somewhat different proposal is to replace older fixed-in-place missiles with ones that are more accurate and more powerful (i.e., each one capable of carrying more warheads). Assuming for the moment that there were actually many fewer rockets and also a smaller total of warheads in the newer rockets, there would seem to be a gain in stability both because the number of new rockets would represent fewer targets and the total number of warheads in the new rockets would represent fewer weapons. However, whatever gain might be achieved in these ways would be lost by making the newer rockets more attractive targets. It has been said by some that it is unrealistic to suppose that the Russians could coordinate a counter-force strike against all (or the vast majority) of the over a thousand US land-based missiles so as to destroy them simultaneously.[30] If that is true, for whatever technological reasons, it might be less unrealistic if roughly the same US nuclear capability were concentrated in a hundred to two hundred missiles. Since the Russians might be more tempted to go after fewer but more powerful missiles than a larger number of less powerful ones, it is clearly important to take both the number of targets (and weapons) and their attractiveness into account if unstable conditions are to be avoided. What is wanted are both fewer and less attractive targets to shoot at. Economic consider-ations aside, it would follow that if the US deployed a fixed-in-place version of its MX missile in addition to those fixed-in-place already, that would count as two steps toward greater instability. If, instead, the US deployed the MX while deactivating its older but more powerful Titan II missiles, and many of its older Minuteman mis-siles, the attractiveness of the new missiles might tempt the Russians to think the unthinkable – especially a few years from now when their command, control and communications systems are better coordin-ated than they are now. So this option still counts as one step toward greater instability.

In theory a missile like the MX could be deployed in a fixed position and yet represent some movement away from instability if a point defense system were placed around these missiles. Economi-cally that might not be the most attractive way to go since the US would have to pay for both new offensive and defensive missile systems. However, if, as its proponents say, ABM technology is both feasible and cost effective, an ABM system can also be used to defend other point targets such as radar installations, naval bases and airports. Unlike rocket-launching facilities, these immovable facili-ties must always be targets given modern technology. But by being defended, they can at least be made less attractive targets. If, again, as the ABM proponents argue, it will shortly be cheaper to defend a relatively small installation than to attack it, the incentive in the next few decades may shift away from automatically targeting everything within reach – which means, of course, targeting everything the enemy has – to targeting only those locations and facilities that are undefendable. Cities might fall in the undefendable category, but more about that later. Naval bases might fall in that category also, since even if seven out of eight warheads aimed at such a base were destroyed, the base itself would also be destroyed. The ships in the harbor would also be destroyed since they could not be moved in the brief time available between the destruction of the first missile fired at the base and the destruction of the base by the eighth missile. In contrast, a military airport might be destroyed by a later missile, but still pose a threat to the attacking enemy since, just before the fatal nuclear weapon hit it, the aircraft based there could have flown the coop. The extra time provided by the defensive missile system before it was overwhelmed might make all the military (and deterrent) difference in the world.

Let us see where we stand at the moment. In the search for the causes of nuclear instability at least five (partly overlapping) factors have already been identified. One factor that seems to cause instability is the sheer numbers of missiles and nuclear bombs deployed. A second factor is the number of warheads in some missiles. As the number of warheads in a missile increases, the missile becomes more attractive as a target. A third factor is the large number of targets, a fourth the accuracy of some rockets (also helping to make them more attractive) and a fifth the total vulnerability of all targets. A sixth and overarching factor is the very nature of the weapons we are talking about. Nuclear weapons cause instability just because they are so powerful. Presumably the proper strategy is to avoid instability by minimizing these and other causes of instability. Unfortunately, one of the complicating things about these factors is that they do not always cause instability. Take accuracy as an example. A very accurate missile might cause instability because it can be used as a first-strike weapon to hit the enemy's control, command and communications centers. But accuracy also tends to encourage weapons designers to make smaller bombs – ones so small that possibly nuclear bombs might no longer be needed to accomplish certain military missions. Similarly, vulnerability is destabilizing in some ways and stabilizing in other ways. Nonetheless, since these factors do on many occasions cause instability, they bear watching and therefore quite properly belong on our list of causes. Although cost does not represent a cause the way the above six factors do, it would also seem to fit into our strategy as something that ought to be diminished as much as possible.

It is in connection with our strategy that the discussion in this chapter has moved to a consideration of the ABM option. And, indeed, there seems to be some utility in developing this option since it would apparently eliminate some targets and make others less attractive. However, there are some formidable objections to it. An often-cited, but in fact not very formidable objection, is that the deployment of an ABM system goes against the spirit of the SALT I agreement. True enough, but that is hardly an objection that should be taken seriously if it could be shown that ABM deployment would significantly lessen the threat of nuclear war. If that could be shown, the proper thing to do, obviously, would be to renegotiate SALT I or let it die quietly. Again, we must remind ourselves that this agreement may have made sense at a time when ABM systems had little chance of working and, further, when these systems themselves had to use nuclear weapons to shoot down nuclear armed rockets. But times have changed and it may be that a modern *non-nuclear* ABM option makes sense now. At least we ought to be open to that possibility.

A more formidable objection is that developing an ABM system represents just another way of spending large quantities of money on weapons systems. The money spent here, the argument continues, would not be restricted to the ABM weapons themselves, but would extend to offensive weapons designed to counter the defensive weapons. Obvious steps that designers of offensive weapons could take would be to make the incoming warheads maneuverable, disguise them to look like space clutter and blind the defensive weapons electronically and physically. So the efforts of the offense to counter the defense, and the latter's efforts to counter the offense, are just two more examples of military escalation. That is the argument.

In reply, we should keep in mind that 'escalation' can be used purely descriptively thus leaving it open whether this particular example of escalation is good or bad. The assumption in the objection is that military escalation must always be bad and, correspondingly, that the only way to move away from unstable conditions in dealing with nuclear weapons is through de-escalation. That is one possible way out, but not necessarily the only way. We may, as we have pointed out already, have to pay our way out by using technology to overcome a problem that technology has created for us. This does not mean that the optimistic claims of ABM advocates about the progress made in this area should be accepted without question. However, it does mean that unless some other objection than high (but reasonable) costs due to technological advances are forthcoming, progress should be carefully watched for what this option might do to help nuclear nations move away from the danger of nuclear war.

Unfortunately, there are still other objections to dealing with the current unstable nuclear condition with defensive systems. One of these is closely related to one that we have dealt with but, nonetheless, differs enough to deserve separate treatment. That objection is that ABM deployment will lead to a final and complete militarization of space since to be fully effective an ABM system must have both a close-in component of small numerous fast and non-nuclear rockets (the endo-atmospheric component) and a Star Wars component of lasers, etc. (the exo-atmospheric component). This layered conception of defense against ICBMs may mean that the fate of nations can and will be decided out there rather than down here. The flippant answer to this objection is that that might not be such a bad idea. Why not fight battles, if battles must be fought, in a place where no one but those participating in them will get hurt? Especially if future battles were fought in deep space, or at worst in orbit, we might assume that the winner in space would control the military situation on earth. Why settle the issue between the west and east in a bloody central front battle with cluster bombs, poison gas and tactical nuclear weapons when it could be settled out there in space cleanly?

If, as we have been assuming, defensive weapons operating both endo-atmospherically and exo-atmospherically come to dominate ICBMs and IRBMs, that dominance has to be complete or almost complete. At least as far as cities are concerned, 90% efficiency is not good enough, as the MAD proponents quite rightly say. Further, offensive missiles will undoubtedly improve in the future.[31] We have discussed these points to some extent already. What needs to be added is that offensive technology itself can easily develop so that it also can utilize the high ground. The offense can play at Star Wars as well by putting nuclear (offensive) weapons into orbit. Having done that, the enemy's reaction time to attack will have been considerably diminished from what it would have been had the attack begun from the ground or the sea. Instead of moving nations away from conditions of instability, it may be that the development of defensive technology will, after all, create new instability. If this is so, the ability to strike swiftly seems to constitute the seventh causal factor leading to instability.

There is another much-talked-about objection to the ABM option that is probably overrated. If, so the objection goes, the promise of the Star Wars defense is fulfilled, there might develop a 'window of vulnerability' in favor of the side that develops this technology first. Having immunity to attack, the Star Wars nation might be tempted to use its missiles before the enemy developed this same immunity. In addition, the other nation might be tempted to strike before the full power of the Star Wars technology could be deployed against it. Thus, in developing a system of defense with a promise of greater stability in the distant future, new conditions of instability might be met in the near future. In effect, this objection shows that rapid technological change by one side can be thought of as an eighth cause of instability.

The reason this problem is probably not terribly serious in this case is that even if the promise is fulfilled, this fulfillment will take such a long time that the superpowers should have ample time to react to each other's research, development and deployment. It is not as if this complicated and very expensive technology can suddenly be put in place to surprise the enemy. Assuming adequate intelligence on both sides, a reasonable assumption in these matters, each side ought to know well enough in advance what the other side is up to. Paradoxically the very high costs of these weapons that lead many to condemn them also serve to guarantee that nations will not be able to surprise one another with an overwhelming technological advance. In this sense high costs serve as a stabilizing force.

With all these difficulties inherent in the ABM option as an alternative to such other options as bilateral and unilateral disarmament, it is important to remember why we took the time to look at this option in the first place. Basically the reason is that in nuclear and other important military matters, the disarmament options have not seemed to work. Nations have little incentive to make them work. These theories of deescalation seem, as a result, less theories than expressions of wishful thinking. One only has to hear the deescalation rhetoric, often expressed in terms of 'If only . . .', as in 'If only we (or all the nations) would do such and such . . .', to realize that these 'theories' would work if nations and people were other than the way they are. Taking nations and people for what they are, we have said that all we can hope for realistically insofar as bilateral agreements are concerned are marginal movements away from the conditions of instability. If nothing else, it seems, ideological commitments run too deep for much more than that to happen. More could happen. To everybody's surprise, the superpowers could agree to cut their nuclear stockpiles by 50%, and then later cut the remainder by two-thirds as George Kennan wishes they would.[32] However, barring such a miracle, it makes sense to continue negotiating for whatever small gains might be possible. It may be, for example, that some agreements will be forthcoming that will forestall the deployment of anti-satellite weapons. By keeping such weapons out of space, the contending nations will at least leave the surveillance satellites up there to do their work of minimizing military surprises. In addition to working to bring about some bilateral agreements, some limited unilateral disarmament steps might also be tried. One or the other of the superpowers could announce that it has decided for a period of a year or so not to deploy this or that weapon and then invite the other side to do the same. If that worked, another unilateral step could be taken.

Still, even if such agreements are forthcoming, it also makes sense not to exclude from our thinking the more expensive options we have been discussing. In this list, deploying mobile missiles, or at least not deploying any more attractive fixed-in-place missiles, seems to be a small step in the right direction. In contrast, defensive systems could change the whole nature of the nuclear threat. Whether that change would be for the better or for the worse, we have concluded, is hard to tell. Yet there is enough promise present that at least research into defensive warfare should be pushed vigorously.[33]

A final point in favor of exploring the defensive option needs to be considered. A defensive shield that makes ballistics missiles completely obsolete may be realizable only in the minds of those who want to manufacture and make profits from such weapons. It may, that is, make no sense to build such weapons systems if the assumption is that they must be either perfect in protecting a nation from a full-scale nuclear attack or be totally useless. But what if the attack is accidental or small in scale? Or what if a rocket comes not from a superpower but from some revenge-seeking maverick nation that has only a small stockpile of such weapons? In today's world of nuclear nakedness, the victim nation would simply have to take the blow and suffer the loss of perhaps hundreds of thousands of people. However, such losses might be avoidable if a not very exotic, endo-atmospheric defense system were in place. As more nations obtain nuclear weapons it might make sense to consider the possibility that defensive systems have a function, even if they are built to satisfy only minimal standards of security.

Tentatively our conclusions are as follows:

1   A minimal ABM system probably should be built. It not only would help prevent casualties in case of an accidental or maverick attack, it also makes certain targets less attractive.

2   Research on ABM systems should continue to help nations determine how viable such systems are. It may be that in the future these non-nuclear systems will make many nuclear weapons obsolete.

3   Fixed-in-place and defenseless ICBMs should be eliminated from military arsenals. These weapons are vulnerable and therefore invite attack.

4   Fixed-in-place *and* accurate missiles should also not be built. These weapons are extremely attractive (especially if they contain many warheads) and, therefore, also invite attack. Since America's Peacemaker missile (MX) falls in this category it would seem to have been badly named by its parents.

5   If new land-based missiles are built, they should be mobile. The principle 'If you can't see 'em, you can't hit 'em' seems to apply here. Missiles that cannot be found are not targets.

6   Bilateral agreements should not be shunned. They probably will yield no more than cosmetic agreements in the foreseeable future but they may do more good than harm if properly handled.

7   Steps to further militarize space should be taken gradually. Since there are few if any disabling agents out there now, it might be possible to come to a bilateral agreement to keep all such agents out of space or, failing that, to keep nuclear weapons down here on earth.

8   Temporary unilateral disarmament should also not be shunned. It probably would do little harm (given the vast array of weapons available) for a nation to unilaterally not deploy (or test) a weapons system until it sees what the competition's response is. However, it would probably not do much good either.

We have deliberately saved one objection to the ABM option for the end of this chapter since it leads directly into the topic of the next one. That objection is that defensive weaponry, especially if it is only partly successful, will make conterforce attacks passé but will do little to improve the situation with respect to a major countervalue attack. That is, such defensive systems will save military installations but not cities and people. As misguided as it is, the notion of a flexible response held by some American officials a generation ago had an element of humanity built into it. That idea was that the US nuclear arsenal ought to be designed in such a way as to be usable flexibly against either cities or military targets. At least that seemed better than simply aiming all US missiles at population centers. As it developed, one misguided aspect of this doctrine was that it encouraged nations to deploy more missiles so that both cities and military targets could be incinerated. Another, and perhaps fatally misguided, aspect of that doctrine was (is) that it assumed that nuclear weapons could be used against military targets, in a counterforce attack, without seriously inviting escalation into a countervalue attack. But now if defensive missiles are deployed, we have seen that they should first and foremost be nestled next to such points of real estate as missile silos and airports. It would be far more difficult to deploy such weapons in all-out defense of large urban areas such as New York City, Los Angeles, Moscow and Leningrad. So defensive weaponry, if successfully deployed, would make the urban areas *the* exclusive targets of nuclear weapons once again. And that raises the question we will now deal with directly: viz., is targeting civilians with nuclear, and non-nuclear, weapons immoral?

·CHAPTER 9

Civilians and the military

I Contributing to the war effort

We speak of nations as being at war with one another. This way of speaking makes more sense with modern wars than it did with wars fought in previous centuries. Governments declare war, and armies, navies and air forces do most of the fighting, at least in conventional wars. But a major modern war is not just the business of the government and the armed forces. As war propaganda on the home front (an interesting expression) declares, 'Everyone has a job to do.' 'Everyone' presumably means 'every adult' and 'has a job to do' is elliptical for something like 'has a job to do to support the war effort.' Even allowing for the propaganda nature of this expression, there is something right about it. The mother who is caring for her own and the neighbor's child makes it possible for the neighbor to work at the steel plant. Her extra effort contributes both to her income and the war effort. The waitress who puts in extra time to keep the cafe open is covering for someone who is serving in the military. However, even if both of these women do other things for the war effort such as buy victory bonds and work as volunteers at military canteens on weekends, they are, at best, part-time contributors to that effort. With or without the war, the first woman would be caring for her child and the second would be putting in her normal hours at the cafe. In contrast, the woman who left her child in the care of another in order to work at the steel plant is making a full-time contribution to the war effort. It is not only a full-time job, it may be more important than the job a draftee is doing painting the barracks 4000 miles from the nearest battle zone.

If contributing to the war effort is a major measure of legitimate vulnerability to attack by the enemy, it would seem that only slackers, anti-war activists, children and the decrepit should get off scot free.

Even military doctors and chaplains should not be totally exempt since the former contribute by making sick and wounded soldiers ready to fight again, and the latter instill soldiers with spiritual calm and confidence that makes them better fighters. It really does seem true that whole nations can be at war.

Nations can be at war also in another sense. Nations are constituted not just by people but also by institutions, and many of these institutions are manifested in buildings and other facilities. To the extent that some of these facilities are used as instruments by those who contribute to the war effort, they too would seem to be subject to attack even though they may not be exclusively military targets. Certain (perhaps all) railroad facilities, airports, bridges and roads, factories, storage facilities and research centers come naturally to mind as being attackable, as well as the equipment (e.g., trains, airplanes) normally associated with these facilities.

These conclusions about people and their material surroundings can be parried by the argument that since people like the baby-sitting mother are only marginally contributing to the war effort, they ought to be exempt from any sort of military attack. Or it might be said of the mother, the doctor and the servant of God that they are serving others *qua* humans, not *qua* military personages so, once again, they ought to be exempt.

No doubt the children the mother is tending deserve total exemption. They certainly do not contribute to their nation's war effort in any way. If anything, insofar as they demand food and clothing that might be in short supply, and take the time of someone who could otherwise be employed in a war factory, they might be said to be helping the other side. In contrast, the mother may deserve some sort of exemption but not necessarily a total one. If, again, the level of one's contribution determines vulnerability, there is something wrong with the argument that she should be totally exempt from attack.

There is something wrong with the *qua* human argument as well. That argument states that farmers serve food to soldiers quite apart from whether they are soldiers. All humans need to eat. Construed narrowly, the argument is that as soldiers, soldiers do not need to eat. What they need are guns and ammunition. Yet, it is not clear why serving soldiers as soldiers needs to be construed so narrowly. Soldiers need to be healthy to fight well and insofar as they are deprived of food for a period of time, they will not be effective as soldiers even if they have guns and ammunition. It is as if the farmer can help the soldier only as a human. The farmer, in fact, is in basically the same position as the baby-sitting mother and the waitress. He supports the war effort by feeding and thereby helping make soldiers strong. But he also is engaged in feeding others, including those who are not contributing to the war effort. Thus he, like the baby-sitting mother and the waitress, is a part-time contributor to the war effort and thus deserves only partial exemption from attack by the enemy.

Later we will see what partial exemption means. For now, however, we need to ask what other measures might get people and things involved in a war. If the nation, not just the government and military, is what is at war, it could be argued that the nation, the whole of it, is attackable. That would suggest that there are no measures of restraint at all. However, we found in Chapter 7, when doing critical thinking on the subject of battle between soldiers, that certain kinds of restraint could be identified. Preferences of everyone concerned arranged themselves such that it was possible to condemn the destruction of military ambulances, hospitals and churches, and the killing of the wounded and prisoners. In effect, critical thinking yielded rules that condemned certain kinds of destruction and killing.

Similarly, even if, in principle, everyone in a nation is targetable since wars can involve whole nations, restraints need not be absent completely. Preferences of people on both sides about people on both sides can be such as to recommend some restraint. For example, we can assume that one of the strongest (rational) preferences of an attacker is to defeat the enemy. We can also assume, as we did in Chapter 1, that people want to live and thus do not want to be made victims of war – whether they are participants or not. For the moment, we can assume that both contributors and non-contributors on the enemy side prefer equally not to be killed (or maimed). Now although their preferences are equal, the attackers' preferences toward them will differ, since they will prefer to disable the contributors more than the non-contributors. That difference in preferences toward those two groups of people goes a long way toward explaining how rules in war can be established for treating people differently. Since the attackers' preferences for hurting non-contributors is weak but the non-contributors' preferences not to be injured is strong, the preference accounting yields the rule that these non-contributors should not be attacked. In contrast, one can more plausibly arrive at a rule about attacking contributors since the attackers' preferences are not as weak with respect to them.

The situation is more complex than this outline of a preference analysis suggests since, in fact, enemy non-contributors to their war effort may wish to avoid being disabled in war more than the contributors. Some within the military have, after all, chosen to engage in activity that is known by them to be dangerous, while others have acquiesced to being drafted. But if preferences actually differ here, that merely gives us another reason for treating people differently in war. That is, it gives us another reason for affirming the rule that those in war are to be treated primarily in accordance with the role they play in it as contributors.

So far, the reintroduction of preferences into the discussion has not brought forth any new measures or criteria to help us decide who is attackable in war. What it has done instead is give us a method that tells us how to discover the measures for which we are looking. By assessing the (rational) preferences of all those affected by a war, we have generated the rule that the less one is involved in (i.e., the smaller the contribution to) a war, the less he is subject to attack.

In Chapter 7 we generated a similar rule by, again, appealing to the preferences of all affected parties. That rule can now be seen to fall under the contribution rule. It says that if a person is helpless, he is exempt from attack. In that earlier chapter, those who are wounded and those captured were said to be included under this general rule. But now in considering the role of civilians, we need to add at least small children and the aged to the list of those covered by this rule. The reason the one rule falls under the other is obvious. Being helpless is as good a reason as can be found for not contributing. Children do not contribute because they cannot. In contrast, their mothers do not contribute because they choose not to.

To see how these two rules function in war, and particularly how they function vis-à-vis civilians, it is necessary to consider them not just in the abstract but to follow their applications in different war settings. Conveniently these settings divide themselves into three types (1) close-in fighting, (2) area fighting and (3) capture (occupation) settings.

II Close-in and area fighting

Consider first the different treatment accorded different people in close-in fighting. The soldiers who come across a baby and the mother caring for it do not, should not, shoot. Nor do they shoot at munitions workers as they see them scurrying for cover, even though they figure that these workers have just stopped making the bullets and shells that have been flying in their direction. Nor do they shoot at those in uniform if they are identified as doctors and chaplains. The only ones they knowingly shoot at are those in and out of uniform carrying guns. This suggests that a third rule is operating to explain why people are singled out for special treatment. It is a rule built upon the other two by specifying the type of participation deserving special treatment. The problem here is that the soldier, the military doctor, the chaplain and the munitions workers are all full-time war participants. Two out of three are even in uniform. However, they differ in that all three pose no close-in danger to the attackers, while the armed fighter does. None of the three is carrying a weapon, none in all likelihood wants to use a weapon, and none can effectively carry on his full-time military duty while the close-in battle is raging. So the general rule is that if the way a person participates in a war endangers the enemy's life, he can be attacked. Specific to close-in fighting the rule is that only those with weapons can be shot at.

Notice how things change when the fighting takes place at a distance and area weapons come into play. First, we no longer can talk about what to do with a mother or a child. The war machine now must crudely identify, not individuals, but groups and, presumably, refrain from using area weapons in those areas where the mothers and the children are not intermingled with military forces. They are exempt from attack for the most basic of reasons. The children are helpless. It is impossible for them to challenge the military forces deployed against them. The mothers and others helping to care for the children have more capability, but probably not enough to handle weapons effectively. Besides they gain immunity by being with the children. Further, if they were to act aggressively against the attacking enemy they might not only lose their lives, but also endanger the lives of the children. So there is little or no reason for anyone to attack children, the women caring for them and older people when they are in camps away from their fighters.

It is otherwise with munitions workers, at least insofar as the factory where they work is still in operation. Before, in close-in battle, they and their factory were no longer contributing to the war effort. If the enemy were moving to occupy the factory, the workers could neither during the battle nor after it be of danger to them. If the enemy were in retreat they could immobilize the factory workers simply by destroying the factory. Beyond that, workers could be taken prisoner if there were a need to immobilize them permanently. But now in area fighting, the plant itself and those who work there pose an almost direct threat to the enemy. What the workers produce in the factory one week can kill a soldier the next week. What one should intend when attacking a munitions factory is the destruction of the factory itself. What loss of life the workers suffer then counts as the unintended (second) but known effect of the attack on the plant. Further, there is an obligation on the attackers' part to keep the loss of life at a minimum so, other things being equal, the plant should be attacked at those times, Saturday night perhaps, when it is practically shut down. However, as we have been arguing, the application of the double (or second) effect principle does not make sense here. The plant and the workers are both legitimate military targets. The plant as an instrument of war is a target without question. But the reason the workers are also legitimate targets goes back to the participation rule. So long as they are participating in war activity dangerous to their enemy and so long as many of them are participating by exhibiting skills that are in short supply, they too are legitimate targets. Far from planning an attack for a Saturday night, morally (and militarily) it would be permissible for the attackers to hit the plant during its peak-production hours. Killing munitions workers in area fighting is no more immoral than killing soldiers in close-in or area fighting. They pose almost a direct threat to their enemy; and since they can be reached only by area attacks (e.g., cannot be captured), their enemy is doing nothing wrong in attacking them in the only way they can.

Thus on our account, war morality is not going to complain if the bombs dropped on a plant destroy the plant and kill its workers inside – or even outside while they are coming to or going from work. War morality would begin to complain if the attack on the workers were carried out in their homes. The complaints would be forthcoming now not because munitions factories are evil and homes sacred. If, in fact, the workers lived together in barracks like regular-army personnel, they would again become legitimate targets. The problem with attacking workers in their homes is that that is where those not participating in the war (e.g., the children and old folks) also live. Besides, there is no need to go searching for the workers in their homes with bombs and rockets since they conveniently assemble daily in their factories. If the intent is to go after those who work in military related industry, there is no excuse for the attackers not to aim their area weapons at the factories where the workers congregate.

Presumably, in addition to munitions workers, those working in electronics, steel, oil refinery, ball bearing, and a host of other plants and facilities could be targeted. All of them in war would be engaged in production which, we can assume, is either exclusively or largely devoted to satisfying military needs. There will, however, be a broad band of workers who are serving the military and the rest of society jointly. Clothing workers and those employed in the food industry are usually cited here although rail workers also deserve mention. The latter, along with their trains, depots, tracks, and bridges have been historically considered worthy military targets. The assumption is that, during war, the rail industry comes to earn this status since it gives the military's needs priority over civilian ones. In a future major war, the airlines will undoubtedly be treated as the railroads have been, and for the same reasons.

The food and clothing industries are a different matter. The food industry is especially a problem, since at the processing, storage, and transportation stages it is concentrated enough, and many of its products are perishable enough, to make it a profitable target for area weaponry. The problem is this. On the one hand, based on the rules concerned with helplessness, the level and type of contribution people make to the war effort, it seems that not all people should be treated alike. Yet, on the other hand, area weapons do not allow the attacking force much leeway. Either they attack to kill (disable) or they do not. Fine-tuning the attack to fit the enemy's condition is very difficult at best. When a food train is spotted by attacking aircraft moving toward, but still 200 miles away from, the front lines, what are the attackers to do? They can certainly figure that some of the food will be dropped off in the cities and towns between the front lines and where the train is located. They can also figure that some of that food will reach the front lines. Further, they can figure that the railroad engine will be pulling tanks, instead of food, in the future and that the rolling-stock can be used in the future to carry military equipment. In theory, once they have done all of their figuring, they could decide to attack the train at a later time after it has discharged its food consigned to the civilians. However, since they or some of their comrades may not see the train again, and since the enemy's train schedules have thoughtlessly not been made public, they must obviously decide to deal with the train where they have found it.

A similar dilemma faces the attackers when they spot a food processing plant or food storage facility several hundred miles from the front line. If anything, the dilemma is more agonizing since an attack here would be on the food chain exclusively, and not also on the largely militarized railway system. Actually, as in the food-train case, there are two separate dilemmas for the attacking pilots because, when they attack the food plant, they are attacking two separate groups. The first is composed of those who work at the plant, the second of those who suffer because the food in the plant is destroyed. The first group is composed of part-timers. They contribute to the war effort but they, even more so, contribute to the process of feeding civilians. To attack them as if they were soldiers is, it seems, an overreaction. If the bombers did anything to these people, it would seem better if they somehow only inconvenienced rather than killed them.

As to those who suffer because the food plant is destroyed, they represent a mixed bag. The attacking pilots want to hurt those who eat and carry guns or, perhaps, those who eat and make guns, but not those who eat and have not, as yet, learned how to speak. But, again, area warfare forces them to either treat the soldiers like civilians (by not bombing their food supplies) or treat civilians like soldiers (by bombing their food supplies).

Before dealing directly with both the train and the food-chain dilemmas, it is helpful to put them into the larger context of a war rather than treat them as acts of omission or commission to be assessed in isolation. Let us suppose that the war in question is a protracted one, and that a nation has blockaded its enemy by land, sea and air. We can imagine a situation much like World War I or II where the pilots are British and the food plants are German. Part of the strategy of the attacking nation is to strangle the enemy economically. Put in this way, the pilots' dilemma has shifted to one that must be dealt with by their political and military leaders.

The question now becomes, are the orders to blockade a country in ways that deprive everyone in it of basic needs such as food (and medical supplies) and, further, encourages the destruction of food trains, food storage, and production facilities moral? In effect, much of what we have already said in this chapter suggests that too much should not be made of such distinctions as between those who are in uniform and those not, those who are combatants and those who are not, and those who are called innocent and those who are not. This is not to say that these distinctions are useless. Those in uniform during war should be treated differently from those not in uniform. And those holding a gun, whether in uniform or not, should certainly be treated differently from those not holding such weapons. But our argument has been that civilians working in munitions factories should also be treated differently from other civilians who just happen to be children or very old. It is not the distinction itself between those in uniform and those not that we are opposing. Rather, our opposition is aimed at any treatment of the distinction as a basic dichotomy so that judgments are made as to when to shoot on an either-or basis. On the account we are opposing, either one is said to have certain rights, duties, etc. because he is in uniform; or have certain other rights, duties, etc. because he is not in uniform.[1] Things in war are much too complicated for such either-or thinking. That is why we are having such a problem with bombing food trains, food processing plants and the rest.

That is also why we are having problems with blockades. A blockade is not an area weapon, but an area strategy. It is, we might say, a dichotomous strategy. Either a nation starts a blockade or it does not. But the blockaded people resist being dichotomized. In area warfare the (moral) attacker wants to avoid a dichotomous response to a non-dichotomous situation, but cannot do so easily.

So what is the attacker to do? Blockade or not? Comparisons are useful here. It will likely be granted by utilitarians and other theorists in ethics that the behavior in the first three items in the list below are as morally reprehensible as can be imagined.[2]

1 Soldiers torture and/or kill infants, children, pregnant mothers, the sick, the wounded, the aged in close-in encounters where it is clear whom they are hurting and it is clear that they are hurting those who are incapable of doing anything militarily.

2 Using area weapons, a military organization attacks refugee camps that contain people of all ages who are clearly not involved in fighting or in any military or semi-military work.

3 Artillery, rockets and/or airplanes are used to destroy population targets such as villages, towns, and cities that contain essentially no military personnel (although they contain some workers who are engaged in military work) and facilities that could not be used for military purposes.

Since the attacks in these three sketches focus upon the wrong people, we tend to think of them as senseless slaughter. A fourth form of behavior, below, is perhaps more controversial since it is aimed not solely at the helpless and totally uninvolved in the war.

4 In a siege, the attacking force refuses to allow the sick and wounded, children, etc. an exit. As a result they suffer starvation and bombardment along with the military and para-military groups in the city.

The strategy in 4, of course, is to keep as many people in the besieged city as possible so as to force those in it to use their food and medical supplies quickly. Part of the brutality of this strategy is to attack those civilians who attempt to find an exit. When that happens, the siege strategy lapses into 1 above. The other part of the strategy is that everyone in the city is attacked militarily by area weapons.

Aside from the hesitation we feel in condemning sieges as quickly as attacks described in 1, 2 and 3, because military personnel are mixed-in with the people, we hesitate for another reason. Those who are besieged have a choice of how to distribute supplies. In truth, it is somewhat misleading to put it that way. Very likely most of those suffering the most will have little to say as to how, for example, the food is to be distributed. Yet in a siege someone other than the attacker has some sort of choice so the burden of responsibility does not fall fully upon the attacking power.

The same is true with a blockade which can be thought of as a loose siege. It is true that the blockading nation creates the situation so it deserves the larger share of the blame. Nonetheless, the blockaded nation has a variety of choices including surrender and various allocation strategies for whatever shortages develop. As it usually does, it can give priority to the military. But it need not. It can put military personnel and children on an equal footing, letting the others fend as best they can. Or it can choose other priorities.

However, just as a siege is at least marginally less severe than attacks falling under points 1, 2 and 3 so, in one way at least, a blockade is less severe than a siege. In and of itself, a blockade need not be accompanied by direct attacks on the enemy that jeopardize the helpless and those otherwise uninvolved in the war. A blockade can be simply a blockade. A siege can also simply be a siege as when the besieger plays a pure waiting game. But once the shooting starts in a siege, it is far more difficult to sort the soldiers from the civilians than in a blockade and, therefore, to keep from attacking the civilian population directly. In contrast, in a blockade, attacks aimed at food trains and the like need not be direct attacks on the population. These people are certainly being attacked indirectly by the blockade and some, as a result, will certainly die. Further, others will suffer in the aftermath by having experienced malnutrition. Yet, overall, because the blockade is not a direct attack and because those blockaded have some choice, a blockade is a more tolerable form of fighting a war than are points 1, 2, 3 and 4.

Another consideration contributes to making a blockade no worse than a tolerable way of fighting. Let us continue assuming that the blockade involves stopping food shipments and even attacking food-supply trains and food facilities within the blockaded nation. It is, in other words, a total blockade. As we have seen, those who have argued that such a blockade is immoral focus on the non-military status of food and food production. Every one needs food, and even soldiers need it, it is alleged, more as humans than as soldiers. Against this argument, we contend that the strength food gives to soldiers makes it a commodity which the nations participating in a war can attack. Depriving the soldier of his food significantly impairs his war-fighting potential. Now if we had drawn a sharp line between soldiers and civilians and said that the former but not the latter can be attacked, it could still be argued that food should not be a blockadable product. After all, more non-soldiers are affected by a food shortage than are soldiers simply because, in most wars, there are more non-soldiers about than soldiers. However, we have not made such a sharp distinction. On the contrary, we have insisted that it is a mistake to make such simplifying distinctions when the situation is not simple. On our account, if it makes sense to blockade food in order to sap the strength of those in uniform, it makes sense to do the same thing to munitions, steel and other workers. In other words, the legitimate target of a blockade is not just those in uniform but all those who are making a significant contribution to the war effort. Bringing all of these people into the calculation of those who can be legitimately harmed makes a blockade less obviously an intolerable military strategy.

That a blockade is at best tolerable can be seen by comparing it with apparently more acceptable forms of military behavior:

A A military organization refuses to use area weapons when it knows that even a few helpless people and those not involved in the war are in the target area. Believing that just war theory dictates that the innocent should not be harmed in war, it attacks with these weapons only when the target area is completely clear of civilians.

B Before an area attack begins, a military organization gives ample warning to the helpless and others not involved in the war to vacate the area. If possible, it even helps them get out of the line of fire. However, after it has given fair warning, it attacks even though it knows there are some people it does not wish to harm in the danger zone.

C A military blockade is initiated. However, it is a blockade with compassion since it permits supplies to enter the blockaded country. These shipments are under the supervision of the Red Cross and other non-partisan relief organizations.

Sandwiching the kind of total blockade we have been concerned with between nos 1, 2 and 3, on the one side, and *A*, *B* and *C*, on the other, makes that kind of blockade look both good and bad at the same time. *C* is especially effective on the side of making the (total) blockade look bad since, to some extent, it attempts to match the variety of people in the blockaded nation with a varied military response. It at least treats some people who are different, differently. What must be asked, of course, is to what extent the humanitarianism of *C* is merely cosmetic. How many helpless people are helped? A token number? If more than that, if, for example, 10% of the people are, how much food and medicine that otherwise would have gone to these victims from the blockaded nation's own stores will now get shunted to military personnel? If enough shunting takes place, the blockade could become ineffective. In all probability because non-partisan groups such as the Red Cross would find it difficult to monitor this situation and, further, would find it difficult to gather enough funds to make a significant dent in the blockade, *C* would not be significantly different from a total blockade. Both *C* and the total blockade would harm a much wider spectrum of the population than would be necessary for hurting the war effort of the enemy but, our argument has been, both strategies, even the total blockade, hurt enough of those involved in the war so that it cannot be condemned. Tolerated yes, condemned or praised, no.

As an aside, it should be noted that tolerance is a concept that has both what might be called external and internal manifestations. Externally other nations and groups can look at what a blockading nation is doing (and how the blockading nation is responding to its priorities) and see both the good and bad in what is happening. Further, they can see that the good and bad are arranged in such a way that tolerance is a proper response. As a part of that response, they will complain a little, if at all, about what is going on. However, they will also monitor the blockade so as to be in position to condemn any shifts away from a close balance of good and bad towards the bad, and praise any shift in the other direction. Internally, that is, within the nation doing the blockading (and within the blockaded nation), a similar monitoring could be going on.

There is a brand of passive or lazy tolerance that does not involve such monitoring. That form of tolerance is akin to resignation. The person or group who tolerates in this sense no longer cares or has been beaten into a position of putting up with some form of behavior. This is not the form of tolerance we are talking about, and our active form should not be confused with it. Like the passive form, the active recognizes that moral judgments, especially in war, do not conveniently sort themselves as either good or bad. Yet the active form also recognizes the danger, in appeals to tolerance, of sliding down the slope by putting up first with this, then that, and finally some other form of behavior. By being active, that is, by constantly monitoring the situation in question, this form of tolerance allows for the possibility that what has been tolerated in the past may no longer be in the future if it verges on getting out of control.

So far in this chapter we have been resisting the suggestion that dichotomous thinking is useful in talking about the enemy. Some of those in enemy country, like children, are incapable of participating in war at all, while others, like mothers, can participate to some degree. Others still, like churchmen, are capable but do not, or they also participate to some degree. Finally, still others participate fully but in different ways. Some in this group make munitions, others use them. Given the varied amount and ways people participate in a war, it makes little sense to treat them all alike (e.g., kill them all) or to try to separate them cleanly into those who are attackable and those who are not. Both in close-in fighting and in area warfare we have tried to show how each group ought to be treated. As it turns out, especially in area warfare, it is not always possible to treat the enemy's people in ways that reflect how they act. The attacker either refrains from attacking at that moment and gives the enemy's fighters and others a break; or he attacks and thereby fails to treat others, such as children, the way they deserve to be treated. Nonetheless, even in area attacks some distinctions are possible, so that those who can be are attacked directly and forcefully, whereas others are attacked only indirectly by means of blockade, that at best is a tolerable way of fighting, or are not attacked at all.

III Occupation during war

Having dealt with how civilians should be treated in close-in fighting and in area warfare, we turn now to how they should be treated in occupation situations. Assuming for the moment that no one in the city or district that has been captured is offering any resistance, it is clear that morally no one should be militarily attacked. So, in that sense, all the distinctions between the amount and type of participation in the war are now irrelevant. But, still, uniforms make a difference. Those captured in uniform, fighters or not, are put in prison camps, while those not in uniform are not. Whereas in area warfare munitions workers and soldiers are treated alike, now they are treated differently. And with good reason. Everyone is disarmed during the period of occupation, but soldiers need to be in prison camps since they are trained to handle weapons and can, if released to live like civilians, take up arms again should the opportunity arise. Aside from being too numerous to be all put in prison camps, civilians, even munitions workers, no longer pose a threat to their occupiers. Uniformed doctors and chaplains are also not dangerous and so could be released. However, since most of them would be needed to do their special work in the prison camps, it should not be considered a great immorality if some are kept in prison along with their fighters.

Of course, enemy civilians, even those who have never been in uniform, are not completely helpless. So an occupying military force can quite rightly place restrictions upon them. The restrictions may even be selective as they would be if a night curfew were placed on all civilians between the ages of 16 to 55. It is interesting here how differently we speak when an army occupies enemy territory as against when it reoccupies some of its own. In the latter case, although we speak of reoccupation as when the western armies reoccupied France in World War II, we do not speak of the Free French and their allies as occupiers or reoccupiers of France. Rather, we call them liberators. These same troops, later, were called occupiers of Germany. This difference in ways of speaking suggests a difference in the way people think that armies should behave when they are holding positions in their own territory and the enemy's. To a certain extent, the enemy's people and one's own have to be treated differently when the immediate battle is over but the war is still continuing. It is true that the enemy's people should not be treated completely differently. Utilitarian (or rights and duty) based theory can easily show the impropriety of, for example, killing enemy civilians and at the same time showing respect for one's own people. Nonetheless, during an occupation, since the enemy's population poses some dangers to the occupier whereas one's own does not, it needs to be treated differently at least until the danger is diminished.

The propriety of treating the enemy and one's own civilians differently in occupation settings raises the interesting question whether the two groups should be treated differently in area fighting when the enemy is still occupying friendly territory. On one theory, they should not be. If there were a war convention that gives rights and lays duties onto people primarily in terms of whether they are in uniform or not, that convention could go on to say that all non-uniformed (innocent) people should be treated alike.[3] Thus, on this account, in air attacks on military targets, the same standards of care used in not harming one's own civilians should be used in not harming the enemy's. Our argument will be that this is not quite right. Just as the two populations ought to be treated differently when an army is an occupier, so should they be treated differently when its enemy is the occupier.

Supererogation, acting above and beyond the call of duty, complicates this whole issue. Let us assume that the attacking air force has commendably high standards about bombing military targets that are mixed in with civilian areas, and that these standards are the same whether the bombers are over enemy-occupied or enemy territory. But let us also assume that on many occasions the crews assigned to bomb enemy-occupied territory take great risks, above and beyond the call of duty, to keep from killing their own civilians. Such bravery may be foolish, but these same crewmen cannot be criticized for not behaving with equal bravery with bombing enemy targets. After all, they are doing their duty. They have reached certain standards set by their superiors for both bombing areas. It is just that in the one case they have chosen to exceed these standards, but not to exceed them in the other.

We suspect that if an air force command tried to enforce an equal standards policy, presumably because civilians are civilians no matter where they are found, that they would not have complete success. Not only would the pilots continue to perform acts above and beyond the call of duty, but they would do so, our suspicion is, in part because of their intuitive sense that the equal-standards policy is wrong.

But are their intuitions correct? To some extent, we can make critical sense of them since we can assume that the occupied factory employees are participating in the enemy's war effort less efficiently than are the enemy's own workers. Presumably they are being coerced into working in these factories. Thus to those bombing the two factories, it will be more important, other things being equal, to harm the latter rather than the former. But there is a far more important consideration at work here. The pilots might put it this way as they approach their target in occupied territory: 'Those are our people down there' or even 'Those are Frenchmen down there.' Part of what they probably mean is that they are fighting the war for the purpose of freeing these people. In that regard, the employees at the occupied munitions factory are totally different from those at the enemy factory. The pilots are not fighting the war for the enemy workers.

The obvious objection to this argument is that although it explains psychologically why the pilots discriminate between the two populations in whether and how they bomb their targets, it is not a moral argument. It is, rather, one not stated within the universalizability principle as moral arguments should be, since it mentions a certain group of people by name. It is an argument involving privilege since Frenchmen (or Americans, British, etc.) are to receive favored treatment. Since one way to state the universalizability principle is that those who are like should be treated alike, and since, in this situation, the two civilian groups are alike in being civilians, it follows that they ought to be treated alike.

The appropriate reply here is that although a people have been mentioned by the pilots in expressing their feelings, the mention is incidental to the pilots' reason for not treating the two populations alike. That reason can be expressed as follows. 'Those people down there have lost their freedom. In a sense it makes no difference who they are. Whoever they are, their own armed forces have an obligation to do what they can to liberate them.' Now this argument does not violate the universalizability principle any more than does 'Parents should care for their children.' It applies to these bomber pilots' behavior but it could have equally applied to the enemy's bomber pilots had their people been suffering an occupation.

So as the pilots approach the target in the occupied territory, they will have mixed feelings. The factory needs to be immobilized and, in a way, so do the people working there. After all their skills are helping to make the munitions as much as the machines in the factory. Yet far from having malice towards these people who are contributing to the enemy's war effort, they want to save them. That is why they are fighting. So what can they do? First, they can refrain from bombing the factory, relying instead on subversion from within to immobilize the facility. Second, if that fails, the factory can be bombed when the fewest number of employees are at work and, third, if it is located near residential areas, a *duty* of greater care can be imposed on the crews not to hit such areas. It is difficult to quantify things at this point but the crews might be expected to take only reasonable care not to hit civilian buildings near the factory in enemy territory, where such care would not entail taking great risks to the air crews. In contrast, the care they would take to avoid bombing the homes of their own people would involve taking some risk – a risk roughly equal to the value of the crew members on the air planes. The point is that the favoured treatment one's own people would receive would no longer be simply a matter of supererogation, but of duty.

IV Nuclear issues again

In the previous chapter we issued a promissory note related to the hostage role that civilians play in this era of nuclear weapons. We are now in position to pay up on this note. From what we have said in this chapter, it should be clear that that role must in some sense be labelled immoral. Although, as we have been insisting, not all civilians are alike and therefore not all should be treated alike, a nuclear threat of attack on all of them instead of on military facilities would seem morally wrong. So also would an actual attack. The trouble is that if the priorities are reversed so that the threat is primarily against military targets (i.e., a counterforce attack) instead of against civilians (i.e., a countervalue attack), the morality of the situation does not seem to improve at all. There is an improvement of sorts in what happens as the result of a first-strike counterforce attack by one of the superpowers. Although it is anybody's guess as to how many people might be killed immediately or almost immediately in a general first strike attack on a civilian population, certainly the figures could be in the hundreds of millions. A counterforce strike of similar dimensions would probably cause civilian casualties in the tens of millions. However, counterforce threats make the enemy more nervous and, as a result, more trigger-happy, so they are thought to be more destabilizing. Worried as the enemy might be that his deterrent could be taken away from him by a preemptive strike, his inclinations to preempt his enemy's preemption increases. Further, if a counterforce attack began, it is not clear, even if it were limited to a certain geographic area (e.g., the sea, the central front), how it could be contained. Further still, many military targets are near civilian areas so that the victim of a counterforce attack might not even perceive what is happening as just such as attack. Mistakenly he might perceive it as an attack on the civilian population. Finally, even if he did perceive what was happening properly, and especially if many of the attacker's ICBM silos were now empty, he might think it best to reply with an attack on the enemy's civilian population. Presumably the enemy's second strike would also be heavily countervalue in nature.

It is possible that a very limited counterforce first strike and a very limited counterforce reply would drive both sides to the peace table once they realized that each meant for its threats to be taken seriously. This is possible, if for no other reason than that deterrance as a policy has lost some credibility as the nuclear arsenals have grown to massive proportions. One side might think that the other side no longer means what it says when it issues nuclear threats. However, the threat of escalation here is considered by most writers to be so great that such very limited counterforce demonstrations of intent are hardly worth the gamble.

The threat of a strike against a population thus seems morally the worst opinion until one considers seriously the implications of a counterforce strike. Then that option seems at least as bad since it would eventually kill the same number of civilians but, in addition, increase the likelihood of turning a threat of nuclear war into a real war. Lamentable as it is, so long as nuclear missile technology dominates the military scene, it seems either impossible or highly improbable that the civilian population can be taken out of hostage. The hostage problem in this sense is identical with the problem of what to do about nuclear-missile technology itself. Thus there is no additional moral problem to be discussed in this chapter that is distinctively different from that discussed in the chapter on weapons with perhaps one exception. That problem has to do with whether there are moral reasons for actually fighting back once a nation has been devastated by a general nuclear attack.

To initiate a first-strike nuclear attack of any kind would seem morally as close to being absolutely forbidden as anything. Even if a nation could get away with it, because it had secretly developed a defensive technology making it completely immune to attack (hardly a real possibility), such an attack would still be close to being absolutely forbidden since it would lead to such a massive slaughter. But what of a second strike attack?

This is not an easy question to deal with from a utilitarian viewpoint since appeals to revenge or to 'right a wrong' are either inappropriate or inadequate. Nor may an appeal such as 'to defend our nation' be applicable since, after the first strike, there would be little or nothing to defend.[4] A measured counterforce response to a measured first strike that is either counterforce or countervalue might make utilitarian sense since, very likely, both sides would be licking their wounds and, as a result, be anxious to come to the peace table. But if the first strike is massive and thus immoral, why would not the response be equally immoral?

It would not be if the second strike were aimed at destroying the government that committed the original atrocity so that it could not commit any additional atrocities on other nations. Any government capable of a first strike that kills millions of people is capable of doing anything. Thus destroying that government makes at least some sense. Unfortunately, in order to destroy the aggressor's government, not only would its officials need to be destroyed, so also would the system of industry, communications, etc. that supports it. In short, the nation as a wielder of power would be destroyed and made such that it would not reemerge as a power in the world.

All this, of course, is insane. No one wants to be in the position of having to ponder seriously whether to fire nuclear missiles after his own nation has been devastated. But if one did find oneself in that position, it is not true that no moral reason could be cited to justify a counterforce retaliatory response. Destroying the aggressor nation might not be reason enough to launch a nuclear reply to a devastating first strike. Such a reply could just be the final blow that would destroy the environment and make life on earth next to impossible for all humans. But let us assume that the victim nation had missiles at sea that survived the first strike action. Let us assume also that in spite of the feverish efforts of the aggressor nation to track down these submarines, they still are at sea and on the prowl. Let us assume further that in time information is sent to the submarine captains that the environment is not in danger if they retaliate. The only danger is to the aggressor nation that now is prepared with its newfound power to dominate the world. Would it be immoral for the submarines to reply to the original attack by firing their missiles? We think it would not be.

CHAPTER 10
Guerrilla warfare

I Guerrilla vs conventional forces

Too much can be said on behalf of the uniform. Military leaders are tempted to treat the uniform as a symbol of establishment power and therefore to show their power by making uniforms as uniform as possible. Even ethicists can find it tempting to sort those in war who can morally be attacked and those who cannot mainly along the lines of those who wear uniforms and those who do not. However, if they succumb to these temptations, these people will find it difficult to deal with guerrilla warfare where one side may not be in uniform at all. They also may be tempted to feel that the guerrillas are not playing the war game according to the moral rules. By not being in uniform the guerrillas seem, so to speak, to be fighting a dirty war.

In earlier chapters we tried to say neither too much nor too little on behalf of the uniform. We argued that the uniform is a useful sorting device. Those in uniform should be treated differently from those not in uniform. And we do not deny that uniforms serve strictly military purposes such as fostering a sense of pride in those who wear them and helping land warriors distinguish between 'us' and 'them.' Still, we argued that the loss of immunity from attack, especially area attack, is not restricted to those in uniform. It is not even restricted to those who have combatant status. Munitions and steel workers, railroaders and even food handlers, among others, all non-combatants and all not in uniform, are subject either to attack or to other forms of military action such as a blockade.

Guerrilla fighters should be seen as just another group of people, uniformed or not, who are subject to some sort of military action.[1] When not in uniform, they differ from (most) other non-uniformed participants in war since they carry and use weapons. As such, unlike the munitions workers, guerrillas can be shot at in close-in fighting. It

might also be argued that they are different from other civilians who
are subject to military action in that they may be deceiving their
enemy with their ordinary civilian clothing, whereas the munitions
workers and other civilians are not. It might also be argued that it is
this deception that adds credence to the notion that guerrillas are
dirty fighters.

The deception is certainly there in many but not all guerrilla wars,
but the deception is not in itself immoral. It is only deception relative
to standards established by establishment powers. For a variety of
military, and perhaps a few moral, reasons, governments follow a
practice concerning uniforms. Following such a practice, there is a
degree of immorality (and illegality) involved when some of their
military people do not fight in their uniforms or switch uniforms for
the purpose of deceiving the enemy. But being obligated to wear a
uniform as the result of standard practice is not the same as being
obligated to accept the practice. There is no obligation that every side
fighting in war must accept any practice pertaining to wearing
uniforms. If there were, and if the obligation were strong enough to
apply to guerrilla fighters, then morality would be serving the
interests of the establishment powers. In many cases at least, for the
guerrillas to put on and continue to wear uniforms would be tanta-
mount to making themselves easy targets for attack.

This point that there is nothing immoral about not wearing a
uniform can be appreciated by reminding ourselves of the danger of
paradigm case arguments. There is nothing wrong with first looking
at war, as we have done, by focusing on establishment inter-nation
wars. That is, there is nothing wrong so long as one does not start
thinking of these kinds of wars as paradigmatic and then supposing
that other kinds such as civil and guerrilla wars are deviant. Wars are
not ordained or intended to be fought in uniform. War is a serious
enough activity to be fought in whatever clothes suit a side's
purposes. When put in that way the point seems obvious, but doing so
reminds us that the immediate purposes of nations who may have
legitimate reasons for engaging in war may not be the same as the
immediate purposes of non-establishment groups who may have
equally legitimate reasons for engaging in war. The guerrillas'
strategy is dictated by the overwhelming strength of their enemy. The
enemy controls the institutions of the society including the banks and
businesses, the public means of transportation, the mass media and
the nation's formal military organization. Again, assuming the

guerrillas have a just cause for going to war, a fight in the open would, for them, be nasty and short.

But the issue of uniforms aside, guerrilla warfare raises a variety of other moral issues that are fruitfully discussed by looking at them from the separate points of view of the guerrillas and their opponents. The issue of targeting, which we have found so vexing when regular military forces employ their area weapons, is particularly interesting because it affects the two sides in a guerrilla war asymmetrically. In a sense there is no real issue for the guerrillas. They know where almost all of their potential targets are. Because the enemy is established in camps, airfields, bridges, radio stations, and other fixed locations, the guerrillas can be as selective as they please. So they need not kill hordes of innocent civilians in the process of achieving their goals. By and large, even their non-state-of-the-art weapons pose few problems. Nor do their hit-and-hide tactics. As we shall see, moral problems in the form of dilemmas can arise for guerrillas but they need not. Because of the position they are in to select their targets, they can fight an especially clean war – if they so choose.

But not so their enemies. Because they cannot easily locate their targets, government forces, either of the left or right, are tempted to lash out at everybody. They will be tempted to search homes, detain people for questioning, arrest suspects, and, further, impose censorship over the press. As for captured prisoners, they will also likely do what Goode describes the Uruguayan government doing to destroy the Tupemaro guerrillas.

> The armed forces revised methods of torture, making them
> more thorough and hence more productive. Captured
> Tupemaros were subjected to imprisonment, blindfolded, and
> were forced to stand long periods of time without food, having
> their heads submerged in water to the point of drowning every
> half hour. Subjected to such torture, the guerrillas talked, and
> the army learned of hideouts, of leaders and plans, and other
> invaluable information. First, the guerrilla *focos* outside of
> Montevideo were eliminated. Then the army concentrated on
> the guerrillas in the capital city. Within six months, the
> organization had been completely destroyed.[2]

Goode is clear that this ruthlessness alone was not what did the Tupemaros in. They themselves had been ruthless (although no more so than many other guerrillas) to the point of losing public support. Still, he says, it was a factor in the Uruguayan case and, insofar as it was, it raises the issue once more about torturing and abusing prisoners. The issue now is somewhat different from that of two nation-states because of the asymmetrical relationship between the two sides. Two nation-states can strike up a convention between themselves to abstain from torturing prisoners if for no other reason than that both are equally vulnerable. One side's people and information can, in theory at least, be equal to the other side's people and information. In guerrilla warfare the establishment is likely to capture and be able to hold more prisoners. Further, guerrilla prisoners will have more information important to the establishment than will establishment prisoners have that is important to the guerrillas. This point is really part of the point made earlier that guerrillas know where their enemy is, but the enemy does not know where they are. The greater need the government has for information, the greater will it be tempted to take short cuts to get it. At this juncture, an argument could be presented that the asymmetrical relationship between the guerrillas and the government dictates that each side follows different moral rules. Thus, if the guerrillas are permitted to have an advantage in not having to wear uniforms, the government should have an advantage in using assorted (one is tempted to say sordid) information-gathering methods.

In assessing this argument, it is well to remember what was said about rules (codes) in Chapter 2. Rules, we said, need to be formulated to help deal with the vast majority of, but not necessarily all, the situations in which we find ourselves. Rules, for Hare and for us, ideally are developed on the critical level but, once developed, serve on the intuitive level to help guide our actions in repeatable situations. Certainly guerrilla warfare is a repeatable phenomenon both in the past and in modern times. So it ought to be possible to develop rules about the treatment of prisoners, suspects, etc. in guerrilla war just as in a conventional war. Whether these rules are different from those in a conventionally fought war, as the above argument suggests they should be, and in what respects they are different, remains to be seen.

Returning briefly to the Uruguayan situation, let us assume that the government's torture methods definitely contributed to the demise of the Tupemaros movement. Let us also assume that the people's preferences in Uruguay were more disposed toward the government than the Tupemaros (as it seems was the case at that time). Given these preferences, we can assume that the people of Uruguay, as a whole, were willing to at least tolerate the government's information-gathering procedures or actually approved of them. More safely we can assume that the Tupemaros' preferences were quite the opposite. But even taking their preferences into account, more people (in and out of the government) probably approved rather than disapproved of the government's ruthless policies.

Nevertheless, such a calculation, crude or refined as it might be, hardly settles the issue of right and wrong on the critical level. Preferences on that level have to be assessed in terms of options. So the question that needs to be asked is whether the Uruguayan government's option is more or less preferred than some other option; that is, are there other and better means available than the one chosen? Consider the situation in Venezuela when the Romulo Betancourt government was faced with the urban guerrillas who called themselves the Movement of the Revolutionary Left (MIR). Although Betancourt by no means used kid gloves in dealing with the MIR, he did use a considerable amount of restraint, waiting, it seems, for the horror of the insurgents' violence to sink into the people's minds before acting. Even after that, he acted with restraint, always being mindful to keep the government's actions more palatable morally than the insurgents'. Edward W. Gude describes what happened next:

> As the elections of December 1963 approached, violence increased rapidly. The insurgents felt that it was necessary either to induce a military coup or to force the government into overreaction before there was a successful transfer of democratic power. Again Betancourt was careful not to overstep his support, and he appeared in public to be responding less forcefully than many would have wished. His was an unstable tightrope to walk, because he was exposed to failure from insufficient action as easily as from excessive action. Because the insurgents were unsuccessful in expanding their scope of operation, the threat against the government did not increase as rapidly as it did in Cuba. A vicious attack on an excursion train in September 1963 provided an opportunity for a very forceful response with large scale popular support. Betancourt took advantage of this

situation to arrest MIR and Communist Party deputies, to use regularly (sic) military troops in urban areas, and to pass emergency measures. These actions were carried out with the strong support of the military and the public at large. As previously, the Betancourt response to insurgent activity appears to have been commensurate with public judgments about what was appropriate.[3]

Gude contrasts Betancourt's restrained policies with Batista's unrestrained ones, and reminds us that, in these cases, the policies of terror and torture did not succeed in the end.

There is certainly no guarantee that a policy of more rather than less restraint will be successful. Indeed, Gude's characterization of Betancourt's policies was in terms of being on 'an unstable tightrope'. Presumably Batista and the Uruguayan governments were similarly teetering. Whatever the case, all we need to establish is whether there is a viable option (i.e., one that has a chance of success) to a Uruguayan-like approach and whether this approach is the one most in accord with the informed preferences of those concerned. Assuming that Betancourt and his people demonstrated the viability of restraint, there is little question which policy, the restrained or the unrestrained, is the most preferred overall. The vast majority of those who preferred the unrestrained policy, no doubt, were found within the government. In contrast, the guerrillas, especially those who would have been tortured under the unrestrained policy, would prefer that the government exhibit restraint (although, no doubt, some guerrillas would favor the unrestrained policy so that the people would have a chance to react against its excesses). The citizens also would favor restraint since they would not want, unless absolutely necessary, to have their homes entered and searched, their freedom of movement and speech restricted, and they or their relatives abused or tortured just because they are suspects.

So there is little doubt what general rule should guide governments in dealing with insurgent groups. Not only is it good moral policy not to torture guerillas, suspects and others, and not abuse the general civilian population when searching for guerrillas, it likely is a good military policy as well, since alienating the population in the process of quelling an insurgency movement plays into the hands of the insurgents.

As always, a general ready-made or intuitive rule such as this one will have exceptions. By exceptions here we do not mean abuses. Abuses are a failure of some kind. An officer or an enlisted man has committed a moral abuse when he tortures someone in violation of a rule that forbids it. An exception, in contrast, is the legitimate overriding of a rule. Here a prisoner is being tortured not merely because he has useful information but, perhaps, because the information he has, but is unwilling to reveal, will save the lives of scores (hundreds, thousands, etc.) of innocent people. He knows, let us say, where bombs are planted that will kill children, those injured, the ill and others.

Earlier we observed that guerrillas are in a special position to fight a morally clean war since their enemies are practically fixed in their gun sights. They do not have to grope around in search of their enemies. Further, they, like the government, have a great interest in maintaining good relations with the populace. Most of Mao's now famous three rules and eight points reflect this interest:

The rules:

1   Obey orders in all your actions.
2   Do not take a single needle or piece of thread from the masses.
3   Turn in everything captured.

The points:

1   Speak politely.
2   Pay fairly for what you buy.
3   Return everything you borrow.
4   Pay for anything you damage.
5   Do not hit or swear at people.
6   Do not damage crops.
7   Do not take liberties with women.
8   Do not ill-treat captives.[4]

He, like most other guerrillas, knew that his forces needed the good will of the people in whose midst they lived and maneuvered. But military strategy aside, the same preference accounting that led us to argue for a governmental policy of restraint applies to guerrillas. The people who find themselves in the midst of a guerrilla war have just as much aversion to being abused by the guerrillas as they do by the government. As to torturing prisoners, the guerrillas have little to gain here and therefore little or no basis upon which to build an argument to justify such activity. In the end, the general rules guiding guerrilla action would seem to be those involving restraint much like those guiding the government activity.

It still might be thought, however, that the moral version of the asymmetry argument might allow guerrillas more leeway in doing things the government ought not to. It is true, as we said, that the guerrillas are in position to fight a cleaner war, but it may be that this observation applies only to certain guerrillas – for example, to those who already have a niche in the society they wish to serve. It might further be thought that since guerrilla warfare is so varied, very likely more so than conventional war between two uniformed military groups, one cannot characterize guerrilla warfare in a general way and then discover general rules that apply to it.

There is some merit to this argument. So-called guerrilla warfare ranges all the way from fighting very similar to that fought between nation-states to military activity initiated by a few individuals. It also ranges from those fought in rural and remote regions to those fought in urban areas. Further it includes war where guerrillas are fighting in friendly territory as against fighting in a country where the population is divided politically, ethnically or religiously. Certainly the tactics of urban guerrillas in a religiously divided country like Northern Ireland are going to have to be different from the tactics of a rural guerrilla in Yugoslavia fighting the Germans during World War II. Granting all this, our problem is to determine whether the morality is going to differ when the conditions of guerrilla wars differ.

II The slippery slope

In a general work on military ethics such as this one, it is impossible to discuss the morality of each form of guerrilla warfare and insurgency. Of necessity, only certain forms and only certain problems found within these forms can be examined. The central problem we have chosen poses a special problem for our own position.

The problem is the slippery slope. We have all along argued against drawing one sharp line separating those who are legitimately subject to some sort of military action from those who are not. Instead we have drawn several lines, none terribly sharp, to help guide the military in various war settings. In the process, we have allowed that a variety of civilians in certain circumstances are subject to attack. The problem is that once we have started down that slope, we may be forced to admit even more civilians in the not-so-charmed circle of those who are attackable. This problem becomes particularly acute since some guerrillas practice what Sir Robert Thompson calls 'selective terrorism'.[5] Such terrorism, he tells us, is perfectly compatible with maintaining a good image with the general population by, for instance, following all of Mao's rules and points. Those selected for terrorist attention are not the people at large but, instead, political and social leaders who are supporting the government – and possibly all those leaders who have not openly declared themselves on the side of the guerrillas. Some of these leaders are attacked and very likely the rest are terrorized. If the mayor of a city is gunned down, candidates to replace him may become a little shy about identifying themselves. Further, once this selective terrorism is applied to other town officials, labor leaders, church leaders and the like, all governmental and social activities associated with the established government could tend to come to standstill.

In addition to the issue of the rightness or wrongness of selective terrorism our problem is whether we have boxed ourselves in so that we are committed to saying that this form of selective terrorism cannot be condemned. Our rules of participation, since they recognize degrees of participation, will make it difficult for us completely to exclude such attacks on political leaders. In World War II, no one would have proclaimed the immorality of enemy attacks on heads of state such as Hitler, Churchill, Roosevelt or even the Japanese emperor. No doubt it would have been both militarily imprudent and immoral if a leader were attacked during a period of the war when he was pursuing peace. Yet during the normal course of the war, these civilian leaders and others such as those in charge of military production (e.g., Albert Speer in Germany), transportation, manpower and resource allocation, are participating directly in the war effort and, therefore, would also seem to be fitting targets of attack. But if a war, especially a guerrilla war, is seen not just as a struggle between military forces, but also as a political struggle, it would seem that all or almost all governmental officials, and not just those participating in the war effort, would be liable to attack. So also would be such institutional leaders as churchmen and labor leaders who, although they are not a part of the government, use their considerable power to encourage others to support the government. In other words, the slope having to do with killing public officials seems to be slippery and also steep enough to take us, if not to a realist bottom, at least very close to it.

Before we deal with this slippery-slope issue, it is necessary to digress in order to discuss terrorism in more detail. Terrorism is, strictly speaking, not a type of war the way conventional and guerrilla wars are. Rather, it is a particular tactic that can be used in either type. During World War II, for example, most of the major powers practiced terror bombing against civilian populations, but they could have chosen not to. It may be more difficult to imagine guerrillas not using terror tactics but we can at least imagine them giving their undivided attention only to strictly military targets.

What gives pause here is that the concept of terror sits uneasily between the actor and the recipient. The actor may intend that his tactics terrorize people, but if he terrorizes no one, we hesitate to call what he has done terrorism. Conversely, he may not intend to terrorize anyone but, inadvertently, succeed in doing so. So even guerrilla attacks on military installations intended primarily to diminish the government's military power can, if the troops are ill trained, terrorize them or others who think they might be attacked in the near future.

Whether correctly identified or not, what is clear is that as a tactic, the intended effect of terrorism is psychological. How that effect is brought about can vary. Screaming dive-bombers, new weapons, massive displays of power, random assassinations, can all have, as the lawyers say, a chilling effect. Morally there need be nothing wrong with this tactic any more than with the tactic of psychologically discouraging the enemy from fighting, by surrounding and then besieging him. In part, what can go morally wrong is targeting the wrong people. There are two, possibly four, different groups that can be targeted. The first is made up of those who are victimized by being killed, tortured, kidnapped, beaten or maimed; and the second of those who seeing what has happened to the first group become terrorized. The third group is composed of those citizens and officials who are harmed because those in the second group act in terror and the fourth is composed of those who are not terrorized but are impressed by the power of the terrorists. Thus the first group is represented by the mayor of a small town, who, along with his wife and children, is killed. The second group is made up of the other officials (and their families) in the town and the mayors and officials in nearby towns, villages and cities. The third and fourth groups would be the people in all these communities who both suffer because their community leaders are unable, through fright, to serve them any longer and who cannot help but think that the terrorists are more powerful than the government. Members of these last two groups might also include officials who, either responding to the terror in the hearts of the people themselves or impressed by it, seek some accommodation with the terrorists.

Taking the acts of killing the mayor and his family as a case study, at least two questions need to be asked. Was it wrong of the terrorists to select those that they did as targets? And, was what they did to them immoral? In terms of the moral standards already established in earlier discussions, some of what they did is uncontroversially wrong. In effect by making the wife and children prisoners, even though they were not involved in the war, and then torturing and finally killing them, the terrorists have done about as much wrong as they could. Arguably the mayor himself is more nearly a legitimate target of attack of some sort. But the treatment accorded him is also unarguably immoral. At worst he should have been taken prisoner and tried in some court if he were guilty of some political, etc. crime. Beyond that, the unarmed mayor was treated much worse than armies were supposed to treat their fully armed and uniformed enemies.

We can get a better sense of what the rights and wrongs of this situation are by asking ourselves what else might the guerrillas have done to the mayor, assuming for the moment that he deserves their attention. He could have been taxed or fined for supporting or being part of the government. Or his office and records could have been destroyed. All of these actions plus the temporary occupation of the town would be more in keeping with those standards of behavior we have already accepted. Those standards, it will be recalled, rested heavily upon a person's level and kind of participation in the war to determine whether and how he is subject to military attack.

But is he morally subject to attack at all? In a sense he is part of the government. This would be particularly true if he were appointed by the central government or, if elected by the people or a local council, he could not assume office without central government's approval. It is considerations such as these that give credence to the guerrilla terrorist charge that the mayor is involved in the war. However, when one looks at the work that a mayor typically does, it becomes clear that his involvement is minimal. Settling most of the town's affairs need not involve him and his helpers with the national government, nor with the guerrillas for that matter. That is, the town needs the mayor no matter who controls the guns in the area in which it is found. So in that sense, he is not really part of the government and is not helping, as a munitions worker is, to support the government or the war effort.

In this sense, had the mayor and his office been subject to an area attack by rockets or mortars that too would have been out of proportion to his involvement in the war and with the government. Area attacks on munitions facilities kill civilians but they kill those civilians and damage those facilities that directly lead to greater enemy firepower at the front. The mayor, at best, is a symbol of government (civil) power and not an instrument of the government's military or political power. Some humiliation of the mayor might thus be tolerable, but annihilation is not.

We can now appreciate better the morality of what is happening to the second group of those involved in guerrilla-terrorist activity. Terrorizing someone is not the same as killing him, but it is serious enough, especially if it is accomplished by a threat to kill that has been backed by another person's death. If the terrorism results from a humiliation of a neighboring mayor or from the destruction of his records or offices, the inclination morally to blame the guerrillas understandably lessens.

In the end, then, our moral judgments of these matters dictate that the guerrillas should abandon some of their terroristic tactics. If the only consistently effective tool of terrorism at their disposal is the killing of officials (and their families) not intimately involved in the war, the moral cost of such a tactic is too high. But certainly other tactics are available for the guerrillas so that whatever chances they have for ultimate victory are probably not seriously compromised by renouncing one kind of terrorism. Aside from those suggestions we have already made concerning what guerrillas can morally do to public officials, nothing we have said thus far forbids other kinds of terrorism. Terrorism practiced against military personnel, for example, has not been condemned. Nor have various power displays such as the destruction of bridges and lines of communication. If these kinds of guerrilla actions convince the population that the central government is ineffective, even though such actions inconvenience and possibly harm them, then so be it. At least these actions exhibit some restraint in focusing on facilities that the government needs to fight the insurgency movement.

So far the asymmetry we talked of at the beginning of this chapter has had less than a devastating effect upon our moral judgments. Although there are differences between the structure of the guerrilla and establishment armies, these differences have made for few differences as to how the two armies should behave. The slippery slope to realism is basically no steeper for guerrilla war than it is for more conventional wars. Just as it is immoral in a conventional war to attack certain civilians so it is similarly immoral to do so in a guerrilla war. In the same vein, if it is immoral to mistreat prisoners in a conventional war, it is similarly immoral to mistreat them in a guerrilla war. In this connection, we have argued that just because the guerrillas do not always wear uniforms is no reason for the establishment armies not to treat them the way they treat enemy uniformed personnel. But guerrillas even more so should not be mistreated when captured since they are fighting in the only way they can. Assuming that they themselves do not commit criminal acts and that they are fighting honorably, they should be treated in the same way as uniformed soldiers when captured.

PART FOUR:
POST-WAR ISSUES

CHAPTER 11
Ending war

1 Goals of war

It is as difficult, and as important, to know when to end wars as to know when to start them. A needlessly long war extends human pain and suffering and can even lead to a reversal of a nation's fortunes. But wars can be too short as well. A war that fails to achieve the goals that justified its start is a waste. A war that ends in the wrong way, on the wrong terms, may sow the seeds of future wars, as World War I did with treaties containing the conditions that sparked World War II. Those who have it in their power to bring wars to an end have the obligation to do so with an eye to the conditions that one ending rather than another will create.

The quick and easy, and unhelpful, answer to these difficulties is that wars should end when victory is achieved or the enemy is defeated. The problem lies in knowing precisely what counts as victory and what as defeat. During the American involvement in Vietnam, the Viet Cong and the North Vietnamese did not win any major battles, nor did they ever place any large segment of the US army in jeopardy. The Vietnamese finally prevailed simply by refusing to be destroyed and waiting for the Americans to withdraw. The Americans did not appear to lose militarily, yet they clearly did not win in any intelligible sense either.

It may also be tempting to think that wars aim at the capitulation of the enemy. The war is over when they are at our mercy. But few wars end with such complete and unconditional surrender. The aims of most wars are satisfied when a piece of territory is exchanged, self-determination is achieved, or reparations made. Seeking complete subjugation could extend the length and intensity of wars far beyond what is necessary for the achievement of most purposes. As we shall claim later, it is rarely necessary or morally justifiable to seek complete and total capitulation of the enemy.

Von Clausewitz offers a helpful clue about when wars should end when he claims that the ultimate goal of warfare is to get the enemy to do our will.[1] The clear implication is that we are not seeking to bend him to any and all of our desires but only to particular ones. We formulate a set of goals or purposes in fighting wars. When we achieve those, there is no longer anything to be gained by continuing the fight. To quit too soon would be to expend our effort in vain while to continue after reasonable goals have been reached is needlessly to waste lives and resources. The difficulty is in determining what goals are justified and translating these goals into specific military and diplomatic aims. Further, as Walzer notes, there is a tendency on the part of governments to inflate the goals for fighting as wars continue.[2] Reasonable and modest war aims are gradually transmuted into grandiose ones. Such goals not only lengthen wars by possibly strengthening the enemy's resolve to continue fighting, but they do not translate easily into specific aims to be achieved by conflict. In addition, such goals may reach beyond what a people can morally justify imposing on another. Having clear and precise guidelines for deciding war goals can avoid this.

A natural way to devise reasonable war goals would be to generate them from the legitimate causes for going to war. If a war were started as a just response to aggression or in self-defense, the goals of war should be to thwart the aggression. Or, if a war were started in response to gross violations of human rights, then the goals of war should be to cause these to cease. Unfortunately, things are not this simple. Let us say that a war has been started in response to unjustified aggression. What should its goals be? Should they be simply to cause the aggression to cease? Should they include the desire to return conditions to what they were before the war began? Should they include reparations, or some other sort of compensation for the sacrifices required for the war? Should they include provisions which will offer reasonable assurances that such aggression will not occur again, such as disarmament, arms control, treaties, etc? Is it reasonable to seek to put national leaders responsible for the wrongful war on trial for their crimes, or simply seek to put them out of office and institute new government? All of these goals can plausibly be said to be generated by the response to wrongful aggression. They range from the relatively modest to the quite ambitious, and from those with which the enemy can agree without great sacrifice to those to which he will agree with the greatest reluctance.

A further complication is that there may be legitimate goals which one may adopt once the fighting has begun but that do not by themselves justify starting a war. Plausibly, an adversary could undertake to correct past injustices committed by the enemy nation, territory wrongfully taken, treaties broken, etc. Or, the liberation of some beleaguered minority might be added to legitimate war aims. An ongoing war may be the occasion to remove a particularly hostile and odious regime. All of these may be legitimate goals to be sought in war though none may suffice to initiate it. So goals may be added that differ from those which were generated by the initiation of the war. Particularly important and difficult goals include creating stability and harmony, or, more broadly, creating a better peace. These could be furthered by readjusting national boundaries, changing governments, improving access to raw materials, creating commissions or agencies to deal with problems, etc.

Lastly, sometimes it will be reasonable to delete goals for which wars were initiated. A war may be started to halt aggression and move enemy troops back to other borders and to neutralize enemy forces as a continuing threat. Once war is under way, the complete set of initial goals may prove impossible to achieve without excessive cost, or the clear-cut military victories necessary to achieve them may prove elusive. Also, things change during wars. Alliances shift, governments change, minority or colonial groups seek liberation or autonomy. All of these can change the picture and make the goals for which the war was initiated impossible to achieve or make it unnecessary to do so.

II Choice of goals

Michael Walzer believes that there are three legitimate goals of war: resistance, restoration, and reasonable assurance that the events which made the war necessary will not recur. In a few, very extreme, cases, such as that of the Nazi government of Germany, Walzer believes it is permissible to seek a complete change of government. He believes that this is the absolute limit of what may legitimately be sought during warfare.[3]

Resistance is simply the effort necessary to cause the enemy to cease his wrongful action, whether it is wrongful aggression or a gross violation of human rights. It is less clear what is included in restoration. Walzer does not discuss this issue, though the possibilities are quite varied, ranging from simple return to pre-war boundaries to restoration of damage caused by war to reparations to help pay the cost of war. Reasonable assurance apparently includes such measures as arms control or disarmament, arbitration or contractual agreements.

There are a number of goals that Walzer explicitly rules out. A nation may not, usually, include as part of its war aims the apprehension and punishment of the enemy leaders.[4] He appears to have two differing arguments to support this. One is his belief that such trials would require complete conquest of the enemy nation. That is, it would require occupying the enemy country and taking over its government. Such an effort would impose substantial costs on one's own forces and would obviously cause great hardships for the people of the nation being invaded. His second, quite different, argument is that a people have a right to their own government, one resulting from the internal political processes of their nation. Just as they have the right to act freely in their individual lives so they have the right to choose their own government. Walzer presumes that a people who have fought in war for a particular government and have made no overt attempt to overthrow it accept it and do not wish to have it altered.[5] To seize their national leaders in order to put them on trial would violate their collective autonomy by overruling their choice of government. This choice, as in the case of the Nazis, can only be overridden in the most extreme circumstances.

Walzer also believes that nations should not be tempted to try to establish a stable legal order and to prevent all international lawlessness.[6] The most that can be sought, he believes, is to respond to each incident as it occurs and undertake war only in response to the most serious violations. This is an interesting claim, but Walzer does not make clear what his reasons are for accepting it. It may be that he believes individual nations simply are not equipped to establish an international legal order. They do not have either the power or the authority to do so. This would be only a practical inability. Or, he might believe that since the only instrument nations have for responding to international lawlessness, war, is so crude and costly, its use cannot be justified in any but the most serious cases. This is another practical problem. A different possibility is that he believes the measures required to insure that lawless acts do not occur will require intrusions into the internal affairs of individual nations that cannot be morally justified. These might include such measures as surveillance of governmental decisions, control of arms movement and procurement, the power to remove governmental leaders for faulty actions or the right to changes in governmental structure or ideological perspective. Walzer's general view that nations have an autonomy derived from the autonomy of their individual citizens would preclude such activity on moral grounds.

Another basic limitation on war aims is that Walzer believes nations should not attempt to impose their ideologies or religions on others.[7] Capitalist nations should not resort to war to convert socialist nations. Democratic nations should not resort to force to overthrow military juntas. Fundamentalist Islamic nations should not go to war to reform their more liberal brethren. Walzer's views here are the result of what Charles Beitz has called the 'morality of states, an international analogue of nineteenth century liberalism,' the view that nations should be free to develop their own ways of life and their own institutions within their own borders.[8] Nations are only justified in intervening to protect themselves or other nations, and they may not use violent coercion to convert others to their own perspective.

The fundamental idea behind Walzer's limitations is that a nation can only make war against other governments. It may never make war against other nations, that is, against the lives of individual people and the social institutions which give these lives meaning and structure.[9] War is an affair involving only governments in Walzer's view. Nations may be harmed only as a secondary consequence of struggles between governments. Furthermore, for Walzer, the only time when the complete destruction of a government is a legitimate goal of war, the international equivalent of capital punishment, is when that government actively seeks the destruction of nations rather than governments.[10] Nazi Germany was the only example of such a government. Japan's government during World War II was not, since its aims were merely straightforwardly imperialistic. It strove to take over nations but not to destroy them as entities.

Unfortunately, Walzer does not explain the basis of his position clearly. It is plausible to assume, however, that it is founded on several considerations. For one thing, only governments are able to do the sorts of things which justify going to war. It follows that wars should be directed against them and, in the extreme case, even seek their destruction. This is so even if we presume that in some sense states are acting in response to the wishes of their own people, for it is the state that must serve as the instrument of these wishes and must therefore be the legitimate object of response. Of course, it is rarely the case that states simply act as instruments of citizens. They can act independently of their citizens' wishes, and, what is more ominous, can shape these wishes to suit their own aims. Furthermore, even if nations could plausibly be said to earnestly support policies that might be said to merit destruction, the remedy is still to institute a change in government. Destruction of a nation is never necessary to remedy even the ills brought about by Nazi Germany.

Finally, Walzer may well argue that the very purpose of instituting rules of war is to protect the nation and thereby the ways of life of its people. He avers elsewhere that the values of the nation are the highest of human life because these provide the framework within which individual human lives can have shape and meaning. To destroy the nation is to destroy the distinctively human life of a people in the same way that killing destroys them physically. It is the sort of argument Aristotle would have understood. Since the purpose of the rules of international conduct is to preserve the nation, its destruction can never be war's purpose, particularly since it is not nations that threaten other nations but only governments.

III Government and nation

Walzer's analysis of the goals of war is sensitive and humane. Also, it is the only sustained and sophisticated analysis presently available. It is not beyond criticism, however. Our analysis of it will focus on some of the tensions among the goals which he advocates. The key to Walzer's position is found in what he says about the relation between government and nation. We will show how flaws in what he says undermine his views on the legitimate goals of war.

Walzer's perspective requires a clear-cut distinction between government and nation since in his view it is at least sometimes legitimate to seek to destroy the former but never the latter. He further presumes, since governmental leaders are sometimes legitimately subject to punishment and even death for crimes of war, that national leaders bear the full moral responsibility for what nation-states do. In contrast, citizens are, at most, passive accomplices. His presumption is illustrated in his view that the leaders of Nazi Germany were legitimately liable to punishment and even death for their acts, while the German people were only guilty to the extent that they did not actively attempt to overthrow the Nazis.[11] They were justly subject to punishment only to the extent of having their political freedom temporarily removed long enough for a new government to be formed.

Unfortunately, Walzer underestimates the complexity of the relationship between government and nation, particularly the extent to which they must cooperate in wartime and the extent to which their fates are intertwined. Furthermore, his attempt to make a clean distinction between the two at this point conflicts with his general view of the close relation between nation and government. Recall that one of his arguments why it is normally wrong to apprehend and try governmental leaders is that government is the result of internal political processes which are a part of the nation itself.

Part of the difficulty lies in the complexity of government as an institution. A government is not simply the set of individuals who happen to be in power in a given nation. Certainly it must include a set of offices with an institutional structure and laws which establish and regulate the relationships between these offices and the nation as a whole. A government may also include the political culture of a nation from which governing practices and institutions of particular governments arise and are supported. Differing relationships among these factors will exist from nation to nation. Sometimes, where a government comes into being as a result of a coup, it may have little relationship to the established culture, and the offices and roles of government will be effectively defined and controlled by one person or set of persons. In such cases to remove these persons is to change and destroy the government, but doing so will do little violence to the political culture of the nation which is either non-existent or was effectively short-circuited by those who took power. In other cases, as in democracies, individuals may be removed from office without harming either the structure and institution of government itself or the political culture from which it derived.

This analysis shows how Walzer is mistaken at several points. It shows that apprehending and punishing leaders need not be equivalent to destroying the institutions of government or political culture. Individuals can be removed with these remaining intact. In cases where particular persons and the structure of government are closely intertwined, this will often be because the individuals in question have shaped this power to suit their own purposes and have little relation to whatever political culture a nation may possess.

Further, it shows that Walzer is flat wrong when he presumes that apprehension and punishment of national leaders requires invasion and occupation of a nation. Nations can, and sometimes do, demand removal of officials from government or their punishment as preconditions for ending wars, returning pieces of territory, or ceasing other activity. Altering the individual persons who are in charge of nations need do little violence to the life of the nation itself.

Seeking to alter the structure of government is difficult enough without invasion and occupation. Certainly the more ambitious goal of altering the political culture of a nation will be even more so without taking these steps. But it is also difficult to fathom what circumstances must obtain for changes in governmental institutions or political culture to be so urgently required that invasion and occupation is justified. Most often the atrocious acts committed by governments are the personal initiatives of the individuals who control them and do not result from structure or culture. It is at least in part because Walzer seems to believe that punishment and trial of government officials requires invasion and occupation and because he does not distinguish between political culture, governmental structure, and the individuals in control of government, that he believes punishing individual leaders is not a legitimate war aim. In fact, the real issue, which he misses, is whether invasion and occupation are ever justified or required in order to achieve one's war aims. They certainly are not justified simply in order to punish errant national leaders, because the cost in human suffering will surely outweigh any benefit that may come from punishing rogue leaders. We will return later to discussion of the circumstances under which invasion and occupation are justified.

Walzer underestimates, as well, the degree to which nation and government are often interconnected in the process of fighting war and in suffering the consequences of the aftermath. Wars are not fought by governments alone. They can only be fought using the resources of the entire nation. They require the support of industry and commerce and often the active participation of individual citizens. Furthermore, whatever consequences result from the aftermath of war will necessarily affect citizens as well as government, whether these include requirements for changes in government, alteration of national boundaries or reparations and indemnifications. It may be that in Nazi Germany the people were simply passive onlookers who could do nothing to alter the course of events. But it need not be this way. In fact, individuals, including many bankers and industrialists, within Nazi Germany *were* active partners of government and some were even tried and punished for war crimes. Walzer is in the uncomfortable position of presuming that governments result from a nation's political culture and that what government officials undertake is their entire responsibility alone. Both of these assumptions may be accurate sometimes, but neither accurately describes all cases. The exploits of Japan during the Second World War, for example, drew on an intense nationalistic and militaristic culture. It would not be appropriate to divorce government from nation in this instance or to claim that the nation is innocent while the government is guilty. Even in Nazi Germany there was substantial popular support for the government. The Nazi movement drew its support from many elements that were deep-seated in German culture. To divorce government and its actions completely from national culture would be mistaken here too. Walzer in fact implicitly acknowledges this in stating that the Germans did require political reeducation following the war.[12]

Where militarism, imperialism and aggression derive from a national political culture, as in the above examples, it is not unreasonable to presume that a legitimate goal of war would be to change the political culture, particularly where national temperament is likely to lead to future instances of aggression. One difficulty with this sort of goal, however, is that it is unlikely to be achieved without invasion and occupation. Thoroughgoing changes in political attitudes and institutions are unlikely to be accomplished without physical occupation. Because invasion and occupation exact a particularly large price in terms of human suffering, they can only be justified where the past aggressive acts are particularly large-scale, widespread and likely to be repeated again in the future. Mere aggression, in and of itself, does not suffice to justify invasion and occupation. Chiseling away at bits of territory or harassing armies or taking over shipping lanes does not suffice – though they may suffice for retaliatory warfare. The costs of invasion can only be justified if the aggression threatens an entire nation, is likely to be repeated, and serves no purpose other than national greed or ideological mandate.

Notice that in the above discussion we mention changing national institutions. Changing national life is not the same as destroying it. Walzer believes that the one action which justifies invasion and occupation for the purpose of changing government is the attempt to destroy nations.[13] In his terminology 'nation-destroying' is the destruction of the way of life of a people and the social institutions which support it. He argues that the Nazis were guilty of this and deserved to have their government removed because of it. The Nazis destroyed nations, straightforwardly, by killing people. There are other examples in recent history of nation-destroying. Both the Communist Chinese following the revolution of 1948 and the Khmer Rouge following their takeover of Cambodia destroyed the nations that they found. They did so not only by killing many people but by tearing down social and political institutions. Their methods, particularly in China, were vastly more successful than those of the Nazis. For the Chinese, the change may have been justified. The nation had suffered a long period of decline. Its social and political institutions were decrepit, and its people were suffering immensely. That is, destruction of the national life of the Chinese people may have been necessary in order to provide the conditions for a better life *for them*. (Whether the particular means the Chinese Communists used to seek these goals, including mass executions and wholesale disruption of individual life, were justified is quite another matter.)

So, in principle, seeking to destroy a nation need not be the grave crime Walzer makes it out to be and may sometimes be morally justified. It may be justified where the character of the nation is such that it is a threat to the lives and well-being of other peoples or where the nation is so decrepit that its own citizens are endangered in both their lives and well-being. In actual practice there are likely to be few if any nations of this sort. The aggressive and militaristic character of the Japanese nation during and preceding World War II clearly caused much suffering, but dealing with the problem did not require the destruction of their nation, only a modification of it. The difficulty is that even substantive modification of national life is likely to require invasion and occupation. While the aims of altering national life may be reasonable in and of themselves, the means required to do so may be too costly.

IV Military and political goals

It would be easy and comforting to be able to say as Walzer does that no goals of war should be sought other than those sufficient to justify going to war in the first place. This may have some effect in limiting the uncontrolled escalation of goals – and the loss of human life and well-being which such excess prompts. Ambitions in excess of those which justify initiating war only serve to prolong it. Walzer's proposal has the advantage of being simple and easily understood. Unfortunately, it is too simple. For one thing, it is not easy to understand just what goals are justified by a given just cause of war. Walzer believes, recall, that just wars have the three goals of resistance, restoration and reasonable assurance that the wrongs which necessitated the war will not recur. These provide little guidance though. Wrongful aggression may justify resistance, but does it justify resistance down to the last soldier? Or, does it always justify seeking a return to the status quo? Or, if it justifies seeking some reasonable assurance that aggression will not be repeated, exactly what measures are allowed? Is it correct, for example, to seek complete destruction of the enemy's military capacity? In other words, we return to our earlier problem that it is not clear just which goals are sanctioned by a justly initiated war. Another complication is that it seems wrong-headed to refuse to add worthwhile goals after a war has begun if they can be achieved at no extra cost or to add goals if they are sufficiently important and beneficial to justify some extra cost.

These difficulties can only be adequately clarified by careful examination of the relation between the political goals sought by war and the military successes required to achieve them. In an overworked quote, von Clausewitz pointed out that the ultimate goals of warfare are always political.[14] The point is that no sane nation engages in warfare for its own sake but that its particular military goals are always in service of some larger political purpose. Achieving military goals places a nation in position to obtain political ones, though they need not do so automatically or simultaneously. Destroying an opponent's naval fleet, for example, may put him in mind to grant a nation sovereignty over a piece of territory or grant mineral rights or freedom of passage in a particular area. The interplay between these goals is complex and intricate.

Straightforwardly military goals are of different varieties. One is actual physical destruction of the enemy's forces, or, more broadly, destruction of his ability to continue fighting. A second, and more elusive goal, is the destruction of his will to fight. Dropping the atomic bomb on Japan did not destroy the capacity of the Japanese to continue fighting but apparently did destroy their will to do so. A distinct set of goals is taking and successfully holding pieces of territory. It may be thought that occupying territory is simultaneously a military and political goal, but this is not so. Occupying territory is not the same as declaring sovereignty over it. One may seize pieces of land which one has no intention of keeping and to which one has no pretense of laying legal claim – both political goals. One may seize territory in order to trade it for other pieces of territory, cut off lines of supply or communication, pave the way for seizing other pieces of territory, or simply because seizing territory is incidental to destroying the enemy's forces. What these three sets of goals have in common is that they are directly achievable by military force. What is essential about them is that they are important only as means, sometimes unsavory means, to achieve other ends. Killing enemy soldiers, for example, or destruction of military equipment is not a good in itself.

The non-military goals sought by warfare are also quite varied. They include rectifying an injustice, recovering a piece of territory, halting immoral actions, increasing security, undergirding world harmony, establishing an international legal system and include also the grander goals of furthering democracy or upholding human dignity. What is important is that, if wars are to be even minimally justified, there must be some plausible connection between the means employed, warfare, and the ends sought.

The distinctive feature of military goals is that they can only be achieved at an unavoidable cost in human life and well-being – or threats to these. In this light, it is worth noting that resistance to wrongful aggression may be unjustified if the military cost outweighs the benefit expected. For example, if an enemy brazenly and wrongfully slices off a piece of uninhabited or sparsely populated territory, military countermeasures may be unjustified if the cost of doing so is great. This does not, of course, mean that no response at all is in order, for diplomatic or economic pressure may legitimately be applied.

In addition to the above, careful attention should be given to the topic of what particular military successes will be necessary to achieve one's goals, and indeed whether military efforts can be expected to achieve them at all. In general, the more grandiose a nation's goals, the less reason there will be to believe that military success of any sort will contribute to achieving them. Careful consideration of issues should include an estimation of what battlefield successes will be necessary to give a nation reasonable assurance of achieving its goals, remembering that *no* battlefield achievements will achieve them directly. Military success only places it in position to make these demands forcefully. Sometimes it may be that war either does not place a nation in position to achieve its goals or that warfare by itself is incapable of achieving them. In some instances even complete capitulation of the enemy – including invasion and occupation – will not suffice to achieve a nation's goals. If Woodrow Wilson was sincere, for example, in believing that World War I was fought to 'make the world safe for democracy', he should have realized that the war was futile, since successful completion of the war was neither necessary nor sufficient to achieve this goal. Wars may, in fact, advance the cause of democracy by defeating aggressors who threaten democratic nations or, following invasion and occupation, by allowing the installation of democratic governments in place of tyrannical regimes. However, in neither case would the cause of serving democracy justify initiating war. In the former case it is the aggression that justifies going to war, not the threat to democracy. If the enemy threatened democracy by spreading an ideology alien to democratic government, war would not be justified. In the latter case, replacing governments is not a goal sufficient to justify going to war.

A final consideration is that the more ambitious a nation's goals the greater incentive the enemy will have to resist them. It can reasonably be expected that the greater the cost to an adversary of yielding to demands, the more he will resist – thus increasing the cost of securing them. A given battle fought to retrieve a piece of wrongfully taken territory becomes much more serious business if it is fought instead to oust a government. For this reason, it is risky to attempt to calculate the probable cost of wars or battles only in terms of the military objectives sought. The political objectives for which military action is undertaken must be weighed into the estimation of the ferocity and determination of the adversary's response. The more ambitious the political goals the greater the military cost is likely to be.

Unfortunately, these considerations indicate that there are no hard and fast simple rules that can be followed in determining what war goals are justified. Leaders must instead pay careful attention to the military cost of various goals and weigh these against any purported benefits. There is no set of war goals which is always and everywhere correct. Some, such as 'the enhancement of human dignity', will always be ruled out as either too vague or simply unsuited for achievement by war. Still others will be justified in only the rarest of circumstances.

V Surrender

The above discussion is based on the premise that wars should only be fought for clear-cut political goals and with a definite understanding of the military achievements necessary to secure them. The implication is that once these goals are within reach, no further expenditure of human pain and well-being is justified, and so the war must end. It seems plausible to suppose that these goals will then form the backbone of some definite set of terms for ending the conflict. Once the enemy is prepared to accept these terms, the war should cease. Note that acceptance of these terms need not be equivalent to surrender or to full capitulation. The terms need not require the forfeiture of the adversary's sovereignty or means of military combat. 'Surrender' implies giving oneself over to the power of the enemy, placing oneself under his control. It seems reasonable to presume that such desperate measures will be taken only in desperate circumstances, when the enemy has lost the means to continue fighting further or when the cost of continuing to do so will be excessive. Surrender in this sense need not be sought as a condition for ending most wars. Indeed, it may be the case that in any but the most important wars, it is wrong to seek complete surrender since the cost here is likely to be greater than ending the war without a surrender. Like General De Gaulle, wise leaders will learn that wars must be abandoned when the cost of continuing them is not worth the further effort.

But there is another essential element to consider. An enemy placed in abject surrender before an opponent is powerless to prevent that opponent from wreaking any destruction it wishes. The consequences of placing a nation in the hands of a brutal, ruthless or even thoughtless enemy can be great. Nations, therefore, have a strong incentive to avoid either capitulation or unconditional surrender. Understandably, they will fight hard to avoid being put in these positions and, as a result, their wars will increase in length and intensity. To avoid this, belligerent nations should state their war aims as clearly and as precisely as possible, so that both sides understand the consequences of defeat, and understand that these consequences are limited. This also gives one side or the other the

opportunity to grant the adversary's demands before they are compelled by complete defeat to do so. Given these considerations, the
question naturally arises whether nations are ever justified in seeking
unconditional surrender from their enemies. If wars are fought for
clear-cut goals, achieving these goals should be given as the terms for
ending war. Seeking unconditional surrender would then be *prima
facie* unjustified.

Michael Walzer appears to endorse the view that the demand for
unconditional surrender is justified in at least certain instances. He
says:

> Concretely, the policy of unconditional surrender implies two
> commitments: first, that the Allies would not negotiate with Nazi
> leaders, would have no dealings with them of any sort . . .;
> second, that no German government would be recognized as
> legitimate and authoritative until the Allies had won the war,
> occupied Germany, and established a new regime. Given the
> character of the existing German government, these
> commitments do not seem to me to represent an excessive
> idealism.[15]

All that seems to be necessary as a condition for unconditional
surrender is the first clause: the refusal to deal with the leaders of a
nation other than to arrange the details of an orderly capitulation.
The requirement of instituting a new government need not be
present. Walzer seems to believe, though he does not explicitly argue
the point, that the demand for unconditional surrender is justified in
conditions where the enemy government is so evil as to lack the moral
legitimacy to be recognized as a government. Seeking unconditional
surrender is a way of refusing to deal with it and thereby refusing to
acknowledge its legitimacy. Seeking unconditional surrender is thus
a way of signalling one's revulsion of an evil regime.

It is difficult to believe that this position is justified, particularly
when its demand is likely to prolong the war and lead to desperate
attempts to avoid final defeat. To begin with, legitimacy is to be
distinguished from sovereignty. 'Sovereignty' refers to the *de facto*
holding of a monopoly of power in a given area. Acknowledging
sovereignty need not imply any particular moral approbation. A
legitimate government is one which has a sound moral claim to
sovereignty. David Luban has pointed out that a great deal of
confusion can be avoided by keeping this distinction in mind.[16] Not

all sovereign governments are also legitimate, but all sovereign governments have a monopoly of power in a given territory. Acknowledgment of sovereignty need not imply acknowledgment of legitimacy. From a moral standpoint, then, no useful purpose is served by refusing to negotiate with a sovereign government. From a practical standpoint, a great deal may be lost by refusing to deal with even a reprehensible government.

The demand for unconditional surrender does not appear to serve any other necessary purpose either. Any group of people, however reprehensible, have the right to know how they are to be treated and to demand certain conditions from their captors. Walzer points out that domestic criminals surrender unconditionally, but their situation is different from nations, since they can be expected to know what sort of system they must confront, what is expected of them and what they can demand after their surrender.[17] Nations do not have such systems of courts or explicit guarantees. Furthermore, authorities sometimes negotiate with criminals when they are in position to do great harm or when harm may be averted by doing so. It is counterproductive, from both a moral and a prudential standpoint, to refuse to deal with adversary nations then if harm can be prevented by doing so.

Walzer, as was pointed out earlier, wants to connect the demand for unconditional surrender to the requirement of invasion and occupation. It is difficult to see why there should be any necessary connection here – or even any natural association between the two. Nations who have determined to invade and occupy others will surely be able to state their demands, and there seems little reason to deny the adversary the opportunity to make counterrequests or demands. In fact, insofar as the conquering nation fails to state in detail what its requirements are, the adversary will have extra incentive to continue fighting as long as possible simply to avoid the uncertainty and anxiety of defeat.

Our discussion in this chapter has shown that it is impossible to make a list of war aims which it is always legitimate to seek. Even resistance to aggression will not be justified in all instances. It is important that the goals of war be weighed against the military actions necessary to achieve them. Some worthwhile ends, such as 'making the world safe for democracy', cannot normally be achieved by war or are so abstract that it is not clear exactly what military efforts will be necessary to achieve them. Wars should be fought with an eye to making future conflict unnecessary. Sometimes, where an intensely military and aggressive culture is responsible for wars, then efforts will legitimately be directed at altering the national culture. In extreme cases, this may require invasion and occupation. In all wars, though, the goals sought and the terms necessary for ending the conflict should be made clear to all parties. We foresee no cases where seeking unconditional surrender is justifiable, though there may be exceptions even to this strong rule.

CHAPTER 12
War crimes and the crime of war

I Two types of crime

If we are to seriously hold to standards of moral conduct in initiating and in fighting wars, we must be prepared to make some response to those who violate these standards. It is hypocritical at best to claim allegiance to particular standards and then stand by idly when they are ignored. It seems reasonable to assume that a response will include punishment and that guilt or innocence must be established by some sort of tribunal before punishment is meted out. Unhappily, a whole host of difficulties arise when these ideas are put into practice.

It is unfortunate that the Nuremburg trials are often taken as a paradigm of what war crimes are like. They are a bad example both because they were conducted by the victors upon the vanquished and in accordance with a set of rules which were created *ex post facto*. They are a poor model as well because they confuse two very different kinds of crimes. Commentators have claimed that the Nazi leaders were on trial for wrongfully engaging in war.[1] This is not the case at all. They were on trial for committing crimes during war. That is, the charges brought against the Nazi leaders were those of torturing, killing innocent civilians, etc., but they were not specifically charged with wrongfully initiating war. As we noted in earlier chapters, there is a difference between *jus ad bellum* (fighting for just or unjust causes) and *jus in bello* (fighting the war correctly).[2] These categories are distinct and need not overlap. One nation may start a war for just cause but use abominable methods of carrying it out, while another nation may wrongfully start war but be a model of propriety in conducting it. Not only are these types of crimes independent of one another, but they are very different sorts of acts.

Individual wrongful acts committed during war are closest to our common-sense notion of what a crime is. That is, they are often committed by single individuals against other single individuals and violate standards of conduct that have close analogues in civilian life. A soldier who wantonly kills an innocent civilian is not acting very differently from a civilian murderer. Both violate the same moral standard since both wrongfully take human life. It is true that the circumstances of the soldier's act may be quite different from those of the civilian. The soldier's conduct may be more understandable and possibly more readily excused than the civilian's act of murder, but the essentials are the same. Wartime cases of massacres, mass murder, or mass torture differ mainly in scale from civilian crimes. We may expect that our common-sense assumptions about crime and punishment will apply fairly readily in these instances.

The crime of wrongfully initiating war is quite different. It does not involve the direct physical assult of one person on another. Neither is there any clear analogue with individual moral conduct. The leader who orders his soldiers to war is not the same as the wife who contracts for the murder of her husband. For one thing, the leader is performing an act that he has authority to perform. National leaders have the authority to go to war, but wives do not have the authority to order the disposal of their husbands. The leader who starts a wrongful war is exercising authority that he *de facto* possesses though he is using it wrongly. As we shall argue later, the leader who starts a war wrongfully has made a judgment which is faulty in some way, such as going to war for the wrong reason or for the wrong ends. Trials of such leaders must then focus on the nature and the circumstances of their judgments. In contrast, war crimes trials must, as do domestic criminal trials, be concerned with events (e.g., when and how the massacre occurred) and the particular role of individuals in these events. Because of these differences, treatment of the issues of wrongful war should be kept distinct from war crimes. This distinction will not always be a sharp one, however, particularly in cases where wars are obviously unjustified.

II The crime of war

In order to see how those who have committed the crime of war should be treated, we must understand just what is wrong about wrongful war and to understand how leaders have acted wrongfully in initiating it. They may have gone wrong in a variety of ways. They may have chosen a means, war, that is inappropriate and inadequate to achieve their ends, such as to gain economic advantage or to aim at grandiose goals (e.g., making the world safe for democracy). Another kind of wrongful war would be one in which the expected benefits are unlikely to outweigh the probable costs, such as going to war to recover an uninhabited slice of previously lost territory. The final sort of wrongful war is one in which war is used as an instrument to gain ends wrong in themselves, such as destroying an unthreatening army or greedily annexing an adjacent nation. Going to war is not wrong in itself, since it is sometimes right and even mandatory. It is going to war for the wrong reasons that makes it an evil. In addition, the wrong engendered by a nation's leaders is not that of ordering people killed. Rather it is creating the risk that they will be. The situation of leaders is not like that when someone orders another person killed but like that when someone orders things done that create great risks for another. It is also important to note that there are presently no clear moral or legal guidelines governing when wars are right and when they are wrong. People do not have the kinds of clear-cut intuitions about them that they have about misconduct on the personal level. This lack of sharp moral insight is reinforced by the widespread view that international affairs is an arena where nations are justified in pursuing their own self-interests, whatever those may be and by whatever means are available. We have argued earlier that this realist position is mistaken and finally self-defeating, but its widespread influence militates against clean judgment in this area. Further, nations are usually clever enough to gloss their actions with some sort of plausible rationalization. Since accurate information is often difficult to uncover in these areas, it is difficult to refute such rationalizations. If we had clear and universally accepted guidelines on these matters, and many hope that we will someday, it would be easier to brand particular wars as obviously wrong and take decisive action against those who start them.

The above discussion allows several conclusions. It is apparent, for one thing, that there are vastly differing degrees of wrongfulness separating those who initiate wars. All cases involve judgments, judgments that often are only shown to be wrong at some future date, when further information becomes available or where it is possible to reconsider in a cool moment. In a few instances it will be possible to say that leaders clearly acted with evil motives and with clear awareness of the nature of their acts. Those who believe that these judgments are easy to make should take note of the incessant scholarly wrangling over the rightness or wrongness of particular governmental decisions. These debates continue even with the benefit of hindsight, new information and lack of the urgency of the moment.

Another conclusion is that those who simply create undue risk are less morally blameworthy than those who intend the ill consequences of their actions. Leaders do not normally wish the death and destruction that accompany wars though they do create the conditions in which these occur. Most of them lack, in other words, straightforwardly criminal intent.

Often, given present circumstances, the actions of leaders who initiate wrongful wars are not sufficiently like criminal acts to warrant apprehending and punishing them. Where war is initiated for clearly wrongful goals, the acts are criminal and should be treated as such. These include cases where masses of innocent people are quite likely to be killed, where fighting is initiated for its own sake, for revenge or to alleviate domestic tensions. All else equal, in these instances, it would be appropriate to consider the capture, trial, and punishment of offending leaders.

Even in these obvious examples other factors will intrude. Often it will be most important simply to end wars as quickly as is possible. Statesmen normally find that this can only be done without attempting to cast blame or make moral judgments about the conduct of the various parties. In a world possessing an institution more closely approximating world government, it may be feasible to always seek to affix blame and exact punishment. We do not have that sort of world at present and may well cause more harm than good by trying to act as though we had. Those who go to war to achieve ends that are wrong in themselves or use war as an inappropriate means to otherwise legitimate ends are guilty of murder, though, as we have argued, they will be guilty in a different way than domestic criminals. In some cases these intentions will be clearly known and so guilt will be readily apparent. Often, though, clear judgments will be difficult to come by. The line between wilful misconduct, erroneous conduct and justified action will be difficult to establish. In the absence of clearly established international standards and recognized international authorities for administering them, there is normally little justification for attempting to try and punish leaders for wrongfully instituting war. The margin for error is simply too great and the possibilities for self-serving action on the part of interested nations loom too large. It is difficult to say, for example, whether the Argentines were wrong to go to war in the Falkland Islands with the British or whether the British were mistaken in responding even if the Argentines were wrong. Even if one side or the other or both were clearly wrong, the situation is simply too nebulous to allow trial or punishment. Also, it is wrong to establish retroactive sanctions. President Hussein of Iraq seems clearly guilty of unwarranted aggression in going to war with Iran in 1981. However, there is no justification for retroactively declaring his acts criminal and then attempting to mete out punishment.

The above does not imply that we should not, in the present, actively make judgments about the moral propriety of initiating various wars. In fact, this ought to be done on a far broader and more systematic basis than in the past. Scholars and jurists should vigorously debate the rightness or wrongness of various wars in explicitly moral terms. International bodies, possibly on the order of Amnesty International, should be established to study these cases and issue reports. This activity can accomplish several things. We would hope that over the years the debates over the morality of various wars would gradually lead to a broad acceptance of a set of standards for rightfully initiating war. We hope that these standards would not be too different from those set out in Chapter 5. What is important is that they gain gradual acceptance, be comprehensive, and be workable. A further, perhaps more important, consequence would be a gradual sensitization of the world community to the view that these are moral issues. Within the United States, for example, debate concerning the morality of various medical practices sensitized many segments of society to the view that there are important moral issues connected with medicine. This had the consequences, not only that people gradually discovered more and more moral problems within medicine, but this sensitivity spread to other arenas, such as business and law. Only after this discussion and sensitization is well developed, over a period of years, can we expect something like a codified system of international law on these matters and the establishment of institutions to apply them. Only when this occurs will it be legitimate to actually carry out trials and exact punishments. We are well aware, of course, that there would have to be additional political and economic developments in the international arena for these features to realized. But we argue that this moral sensitization and debate is a necessary, if not a sufficient, condition for such developments.

III Responsibility for wrongful war

If we imagine for a moment that the time has come for establishing judicial proceedings on the international level, what should they be like? The unstated presumption of this chapter has been that national leaders will be held both accountable and punishable as individuals for whatever misdeeds they undertake while in office. International trials and punishments thus will be focused on individual leaders.

In any government, however democratic, there will always be a determinate number of officials who have primary responsibility for formulating any given governmental action. They will be responsible for these actions in the sense that they possess both the power and the authority to order them. Even though they are only able to undertake these actions in virtue of the role which they play within the governmental structure, they, as individuals, are the agents who initiate them. Therefore, the same account of moral responsibility that applies to individual action can be used with minimal modification to characterize the actions of individuals when they act as governmental officials. This analysis would also apply to large collective bodies of individuals who enact governmental directives as a unit. In cases, such as that of the United States, where the legislative and executive branches are distinct and possess different and interlocking authority in the matter of the declaration of war, the analysis would be much more complex and the result much less clear-cut.

Given that there are governmental leaders who are the agents of wrongful war and given that reasonably exact standards could be established for determining the rightness or wrongness of wars, there is no reason why the model of individual criminal trial and punishment could not be used. The criminal model would be appropriate because it would be assumed that governmental officials knowingly took actions that caused the wrongful death of persons or destruction of their society. Punishments could include simple fines, removal from office, imprisonment or even death depending on the extent of the wrongness of their actions and the extent of their responsibility for them. Severity of punishment would depend not just on the numbers of people killed or amount of property destroyed but on the extent to which these fail to be justified in the service of any worth-while goals which might significantly benefit individual persons.

It may be argued in response that it is unfair to hold political leaders to account in this fashion. Some may claim that leaders are not seen as criminals when they adopt disastrous policies in other areas, so it is unfair to punish them personally if they go to war wrongfully. This argument ignores the distinction between making a difficult but wrong choice and knowingly embarking on a wrongful course of action. At the point in time at which criminal punishment would be justified, there would be clear-cut standards for determining rightful from wrongful war and there would still be, as there is in criminal law, allowance for honest mistakes. The analogy with individual police officers would apply here. Police officers are sometimes allowed to use deadly force in carrying out their responsibilities. They also are forced to make decisions about when or when not to kill, and must do so under difficult circumstances, without full information, and under emotional stress. But they can still be tried and punished for murder if they blatantly ignore the standards which govern the use of deadly force or make inexcusable errors in judgment. Political leaders should be held accountable in similar fashion. Once a system of standards is in place, political leaders would be expected to know what they are and to conduct themselves accordingly.

It might also be objected that political leaders have the primary responsibility of serving their nation's interest and are therefore justified in going to war if they feel the good of the nation requires it. It is true that some wrongful wars violate even *this* standard, when leaders initiate conflict for their own personal benefit. So the standard does have some bite, but there are reasons for thinking a stronger standard is necessary. To begin with the obvious point, the standards of just war are meant to benefit all nations and all peoples, including the leaders in question. Because of this, it will be to their long-term advantage to see that the standards are observed by all. Further, while leaders are, as we indicated in the Introduction, justified in giving greater weight to the interests of their own people, they are not justified in counting all the interests of their nation as greater than any of the interests of any other nation. When it is a matter of life and death, they will be justified in putting their own citizens first, but they are not justified in placing minor interests of their own people ahead of the vital interests of others. So leaders may serve national interests, but they are not justified in seeking out any means whatsoever to do so.

In sum, there is no reason to fail to hold political leaders accountable for wrongful war in the same fashion we would hold them accountable for common criminal acts committed while in office. If these ideas seem odd and discordant with our usual moral sensibility, it is because we are not used to judging decisions of policy in this fashion and we do not as yet have the institutions necessary to put these judgments into effect. There is no reason to believe that making judgments of this sort and acting on them will seem strange once the transition period is past.

One response to this argument is that governmental decisions are too complex, since they derive from too many sources, for us to be able to pin the blame fairly and squarely on one or a few individuals. A quick answer to this complaint is that very often wars are the direct initiative of a few individuals who clearly hold the reins of power. Even where governmental power is more diffuse, as in the United States or the nations of Western Europe, decisions as important as starting a war are channelled through clear-cut lines of authority. Since no nation wants to go to war by inadvertence or accident, it will normally be very easy to determine those who are responsible for a wrongful war.

There is precedent, on the level of ordinary criminal action, for holding a variety of people culpable for a highly complex, well-organized criminal act, such as robbery or a murder plot. In these cases guilt and punishment are assigned in accordance with each individual's role in carrying out the act. There are difficulties even on this level, of course. Different jurisdictions assign different levels of punishment for accomplices, drivers of get-away cars, for example. In these marginal cases ethicists would do well to take a cue from domestic law. Law makers do not fret about devising the one right answer to these problems. Rather they devise a workable answer that is then promulgated so that all are aware of how they will fare if they play one role or another in criminal acts. If the legal determination of how to apportion guilt to the get-away car driver turns out to be unworkable or unacceptable, it can always be altered at a later date. Ethicists sometimes worry too much about devising a precisely correct answer in marginal cases where no such precision is to be had. Even in complex governmental actions involving the agency of hundreds of people, there is no reason to refrain from holding them all accountable to the degree of their agency in carrying out the wrongful act.

IV Trials of crimes of wrongful war

If trials are to be held, the lesson of Nuremberg is that they should not be conducted by the victors and that the procedures and standards should not be devised on an *ad hoc* basis. Victors in war are simply too powerful and too likely to be biased and vengeful. Furthermore, simple justice requires that the structure of the courts, their procedures and standards of judgment should be instituted in peacetime, long before they are likely to be put to use in any particular case.

There is no reason in principle why such courts could not be part of the judicial system of the nations whose leaders are charged with the crime of wrongful war, provided such courts could be made sufficiently independent. In fact, there may be good reasons for adopting this approach simply because the nations of the world are more likely to agree to such a system if the courts are at home. Also nations may be more disposed actually to allow cases to come to trial if handled by national courts. The obvious difficulties of such a system, involving lack of a single standard of judgment and the tendency of such courts to be subject to governmental manipulation, could be met by having international review boards set standards and review the conditions under which courts operate. Another point in favor of national courts is that the international courts have enjoyed, at best, mediocre success, functioning only when both litigants feel they have an advantage in turning to them and in being generally toothless. It may be, of course, that the problems with establishing truly independent national courts will be so great that only an international one will be able to provide any semblance of justice. In that case they should be tried as a second-best option.

Whatever system of courts is established, it will be essential for their proper functioning that they have adequate evidence concerning any given case. In fact, short of the barriers to actually bringing cases to trial, problems of getting adequate evidence are likely to be the most difficult the courts will face. They seem to require information not only about military movements but about the inner workings of governments. If the courts are genuinely sovereign bodies, with powers to subpoena witnesses and documents, these problems would be mitigated, though they would still be substantial.

With just war issues, however, these problems are less imposing than they may at first appear because the problem of the courts will not be so much to decide who did exactly what, but whether a given nation was justified in entering a given war and just who was responsible for making the decisions to go to war. So, in contrast to criminal courts, they will have the task of applying standards of just war to particular cases to determine whether the decisions of particular leaders were justified. Making these judgments does not require intimate knowledge of the workings of a particular government or esoteric information about a given conflict. The difficulty can be met in part by requiring nations wishing to declare war to justify their decision by using a standard format created by international conference. Leaders would be forced to explain in detail their reasons and motives for going to war and supply documents to support their contention. These documents could then be used as the basis of judicial proceedings in cases where wrongful war is charged. What is essential is that the machinery for gathering these documents be in place permanently and that standard formats for justifying the decision to initiate war be agreed upon long before war starts.

The obvious loophole here is when nations do not declare war but simply slide into it as the United States did in Vietnam and is in danger of doing in Central America. The way to meet this problem is to rely on the criteria of Chapters 1 and 5 to determine when nations are or are not at war. It should have been possible to say, after a certain level of commitment of troops and active engagement in conflict, that the United States was in fact at war in Vietnam whether it acknowledged the situation or not.

There remains, of course, the problem that the leaders of nations are unlikely to deliver themselves up for trial. Even those nations that lose wars, but are not invaded and occupied, are unlikely to be in such desperate circumstances as to make their leaders readily available for trial. Fortunately, leaders need not be physically brought before the court for its decisions to have great impact. All nations wish to enjoy the respect of other nations and to avoid the status of international pariahs. A leader found guilty of wrongful war would lose a great deal of prestige and respect both at home and abroad. Beyond this, there are more forceful measures that the international community could take to punish such leaders. The most obvious would be international ostracism. Nations could withdraw diplomatic recognition.

International organizations, such as the United Nations, could suspend membership and assistance. Nations could stage economic boycotts until the convicted leaders were punished. None of these steps requires invasion and occupation. None requires international violence, but all are likely to be effective if supported by the international community. A further point is that it should be made as easy as possible for nations to comply with demands for punishments of convicted leaders. One way of doing this would be to allow these nations to carry out punishment themselves. There is no reason why convicted leaders could not be punished by their own governments.

V Punishment of nations

If those leaders responsible for wrongful war are punished as individuals, it makes sense to ask whether any further punitive measures against the errant nation as a whole are in order. It makes sense because, as we pointed out in Chapter 11, the nation as a whole must actively take part in undertaking a modern war. The usual way of punishing a nation in the modern era is to demand punitive reparations. Other measures might include depriving it of certain rights pertaining to trade or the right to maintain armed forces. Or nations might pay the price of losing political autonomy or being dismembered and parcelled out to other nations.

Certainly many, perhaps most, people within a nation would support its wars and even benefit from them. Being in favor of imperialistic warfare is wrong, but is not the same as being an agent in bringing it about. There is a difference between someone who applauds the bravado of a bank robber and someone who drives the get-away car. Only the latter is an accomplice in the commission of crime, and even the driver is often thought to be deserving less severe punishment than the robber.

The great bulk of citizens who support wrongful war are not accomplices in carrying it out and so will not be appropriately liable to punishment. Furthermore, ordinary citizens normally do not have sufficient information to be able to judge whether a given war is just or not. They are also subjected to considerable pressure from all sides to support their country blindly, particularly during war.

It might be argued in response that ordinary citizens are instrumental in wrongful wars since they serve to operate the institutions, including the military, that are necessary for executing any war.

This view is mistaken since these citizens serve only as the instruments through which these institutions function. Further, the institutions themselves, and those who lead them, are instruments only as far as governmental policy is concerned. Only governmental leaders themselves are genuine agents in these cases. To acknowledge, as we do in Chapter 11, that nations may sometimes need to be altered so as to prevent future aggression does not imply that the citizens whose lives are thus changed are guilty in the usual sense of the terms or that such alteration should be viewed as punishment.

The above applies to military leaders as well. The proper function of the military is to serve as the instrument of governmental policy. Aside from a few exceptional cases, mostly in South America and Africa, this is the function they do have in most nations. Nonetheless, military leaders, because they are in direct control of the instruments of violence, are uniquely well-placed to halt wrongful war. However, it is very difficult and unwise to make provision for this judgment as part of the normal mechanism of the military because of the dangers of politicization and loss of civilian control. Such judgments, if they are made at all, would have to be the individual, and private, judgments of well-placed officers who could then carry them out at their own risk. We may well believe that individuals in certain cases have the responsibility of doing just this, and we may sometimes legitimately censure them if they do not. But it is wrong to hold them personally liable and accountable for punishment if they do not. The reason, once more, is that they are simply not the agents of these decisions.

Some may be impressed with an argument Walzer makes. He claims that sometimes a nation will be legitimately accountable for an obviously wrongful war when it fails to overthrow the government conducting it. He then argues that they are punishable only to the extent that they lose their right to govern themselves until such a time as a new, more legitimate, government can be formed for them.[3]

Walzer seriously underestimates the difficulty of overthrowing a stable government, particularly in wartime where everything is keyed to pursuing the struggle. It is hard to see how it could be a duty of citizens to rise up in such circumstances and even harder to see how they could legitimately be subject to punishment for failing to do so. Possibly Walzer is confused by the example that he uses to illustrate his point and because he has come to the correct conclusion for the wrong reason. He believes that it was legitimate for the German people, following World War II, to lose their right to govern long enough for a new democratic government to be fashioned for them. He is correct in this view, but wrong in believing the actions are justified as punishment and thus count as intentional harm caused to the people of Germany. For one thing it is not obvious that having a democratic regime fashioned for them can be viewed as intentional harm. But even if it is, it is justified only as a measure to prevent future aggression by Germany and not as punishment. If the Germans had shown signs of being able spontaneously to form their own non-imperialistic and non-militaristic government following the war, there would be no justification for seeking to deprive them in some other way. So imposing a new government on the Germans was justified but only as a measure to forestall future aggresssion and not as punishment.

VI Reparations

It is possible, however, that though punishment of nations is not justified, demanding reparations from them will be. On the domestic model of tort law, they could be viewed as just compensation for harm done. Wars, after all, cause huge amounts of human pain and suffering, as well as the wholesale destruction of the means people require to live decent lives. It is legitimate to inquire whether the citizens and governments of the nations who have instituted wrongful war should be required to be of assistance in attempting to repair the damage done. On the model of the domestic law of torts, such mandatory payments will not be viewed as punishment and those who are liable to bear the burden of paying them need not be viewed as guilty. An individual whose tree falls on another's garage may be held liable for damages just because it is his tree, and not because he is somehow at fault for the incident.

If it seems right that those who suffer as the result of unjust war should receive some assistance in repairing the damage of war, it also seems right that those who are part of the nation in the wrong should have a duty to make payments. The major, interlinking issues are those of how large the payments should be and what purposes they should serve. A number of observations will help focus the discussion. Since the reparations are not justified as punishment, they should not be so large as to seriously impair the lives or well-being of the nation required to pay them. The payment should not impair the normal functioning of the economy and should not interfere in any serious way with whatever rebuilding effort the nation may itself have to make following the war. Neither should the payments be such that they will hold the nation in permanent economic bondage to the recipient. If the guilty nation is itself devastated by the war or is simply too poor to afford reparations, they may either be cancelled or deferred until such a time as it is able to pay. Or, preferably, some token reparation may be required as a gesture of common humanity and acknowledgment of responsibility for the harm done.

Another point is that it will obviously be impossible to repay anywhere near an amount adequate for all the physical and emotional damage done by war. There is no way to put an adequate cost on lives lost, years wasted, fears undergone, or environment destroyed. Attempts to compensate in this fashion will necessarily be inadequate and will likely be so huge as to be punitive. For the same reasons it will be impossible to return conditions to their pre-war state. The burden would be overwhelming and would, in any case, be impossible to achieve.

It is well to recall, in addition, that some wars may be so small in extent or so isolated geographically that they cause no significant harm to individual life and well-being. Sometimes the war may not even impose any significant financial burden on the countries who fight them. In these cases, where no significant harm is done, there is little justification to the demand for reparations. In yet other instances the victimized nation may be so wealthy or have such an abundance of resources that it can well afford to repair the damage done, and again there is little justification to the demand for more than token reparations.

In light of these considerations, it would seem that the point of reparations is to assist the individuals who have been harmed by the conflict to return to living as near normal lives as is feasible. The focus must be on individual citizens because, as we argue in the Introduction, they are ultimately all that count in international relations. What may be harmed by war is their physical existence and their ability to live in minimally adequate fashion. So reparations should not be directed towards governments or other institutions except insofar as this will help individual persons. Neither can reparations be expected to do the impossible. They cannot return the dead to life nor can they resurrect what has been destroyed. What is possible, and feasible, is that those who have been harmed be assisted to live normal lives or, where this is impossible, to lives as near normal as possible. It is of interest that, given these goals, reparations need not be financial, nor need they include material goods. They may take the form of human services, transmission of expertise, formation of cooperative ventures, or the establishment of special trade relations. All may serve the purpose of a return to normal life. In fact, many of these measures may be preferable because the bonds of common humanity they foster may serve to increase mutual understanding and perhaps forestall future conflict.

VII War crimes

Crimes committed during the course of war, war crimes, differ in several ways from the crime of wrongfully initiating war. For one thing, war crimes have been the subject of much greater international scrutiny. As a result, there are a variety of commonly accepted standards of conduct during war and a number of international conventions concerning them. War crimes, more than the crime of wrongful war, fit our usual ideas of criminal conduct.

Another difference is the diversity of types of crime. One is the crime of individual against individual, the solitary soldier and the enemy civilian family. A second level is the wrongful command, where an individual is giving orders that will be carried out remotely by others, such as the order to destroy an enemy village. These sorts of crime differ also in complexity, involving numbers of people with varying degrees of responsibility for the final outcome. Lastly, there are the crimes of policy or strategy. In these crimes, the agency is not one of direct command but of general directive, such as the decision to use area bombing rather than precision bombing or the decision to use biological warfare or famine to reduce the enemy. Strategy does not initiate a single set of actions but a range of responses. It sets guidelines for action and goals to be achieved rather than particular sets of orders. Furthermore, the agency will be of the most abstract and remote sort, often issuing from committees perhaps or from a variety of sources in the chain of command.

There is some debate whether it makes sense to attempt to punish war crimes at all. One strand of the debate is that individuals, because of the pressures of circumstance and the desperate nature of war, are not fully in control of what they do. There are really two arguments here, directed against differing levels of activity. One is that the ordinary soldier, in the stress of battle, cannot be expected to be held accountable for what he does. The second is that in war nations can only have winning as a goal, and they cannot be expected to forgo any measures that will help them achieve that goal.

We have responded to both these claims earlier. Soldiers are expected to obey orders and are punished if they do not. If they are sufficiently in control of themselves to follow commands, and no army could function if they could not, there is no reason to believe that they will be unable to follow moral imperatives as well. In addition, insofar as training and the institutional focus of the military shifts in the future to a greater emphasis on moral accountability, some sort of system of punishment and reward will become even more important than it now is. In response to the second argument, wars are rarely literally a matter of life or death for nations – 'serious wars' in Richard Brandt's phrase.[4] There are a large number of wars going on all the time, and they are rarely of the cataclysmic sort. In a few, rare, cases adopting otherwise immoral strategies may possibly make the difference between the survival of a nation and its destruction. In these cases such means will be justified, as we argued earlier. In any case, once a nation adopts a certain strategy, the enemy is likely to respond in kind, and so the advantage will be lost and the war will become more savage to no good purpose. Lastly, the great majority of likely war atrocities, such as the slaughter of civilians, are unlikely to confer any substantial military benefit on the perpetrator.

Another strand in the debate about the feasibility of war crimes trials is concerned with whether punishment for war crimes can be expected to deter future transgressions. The claim is, generally, that instances of war crimes are so rare, the circumstances of their commission so varied and trial and punishment so erratic that it makes little sense to claim such trials have a deterrent effect.[5] It is hard to know how seriously to take some of these points. Proponents of these arguments seem to overlook just how common wars are and how frequently atrocities occur within them. We have pointed out earlier that nearly a quarter of the nations of the world are presently engaged in one form of war or other. Most of these receive little coverage in the press. Nonetheless, we hear about atrocities even in these wars frequently enough. As for the claim that the circumstances of fighting are quite varied, there is something to it. There are conventional wars, civil wars, wars of terror, and guerrilla wars; and they are fought for different reasons in different cultures. But the variations are not so great that these wars do not fall into a manageable number of categories. It is true, for example, that shooting an unarmed civilian may be easier to justify or easier to understand in guerrilla war than in conventional war. This implies only that the cases will be handled differently, just as murders in ordinary life occur under varied conditions, and these differences are taken account of in judicial proceedings. The last point, that trial and punishment of war crimes is quite erratic, is both true and important. Nations have quite differing institutional machinery for handling war crimes and some take these proceedings far more seriously than others. Even in the most conscientious of nations, very probably few of the war crimes committed are discovered or brought to trial. So there is unlikely to be a significant deterrent effect.

The proper response to this difficulty is not simply to ignore war crimes, but to work hard to develop the kinds of attitudes and institutions in which there will be adequate enforcement of them. The present way of doing things in which there is little or no serious attempt to manage war crimes or in which war crimes trials are undertaken by the victors once the war is over is grossly inadequate – and perpetrates a good deal of injustice. Only a small fraction of the murders in the United States result in trial and only a smaller fraction result in punishment. But people do not conclude from that that it is useless to attempt to enforce laws against murder. Rather, they urge that the system for handling these cases be made more effective. The same sort of conclusion applies to war crimes. Present inadequacy is not a reason for doing nothing. It is a reason for reforming and improving the system. Contrary to the impression given by the Nuremberg trials, the main responsibility for controlling crimes of war should lie with individual nations. Response to such crimes should be made during the war itself, as other kinds of criminal activity, rather than awaiting a final accounting at war's end. Machinery for handling war crimes should be part of the judicial mechanism of the military of each individual nation. Of course, there would need to be agreement on an international basis well before wars begin on standards for handling cases of war crimes, making charges against perpetrators and assigning punishment. In contrast to the lack of clarity about what constitutes war, there is broad international agreement about what actions count as war crimes. These understandings need to be refined and elaborated to become fully workable. We would hope that eventually there would be an international body of case law which would further refine our responses to war crimes.

The problem, of course, is that different nations will take these crimes more or less seriously. Further, often the victims of these crimes will be the enemy soldiers or civilians, who will not be well placed to make complaints to the appropriate military authorities. For these reasons we would, once more, recommend a two-tier system for dealing with war crimes. Primary responsibility for dealing with them, including actual trial and punishment, would lie with the military powers themselves. But these should be supplemented and reviewed by international judicial bodies who would monitor trial procedures, the handling of complaints and punishments. Furthermore, these bodies would have to establish liaison and investigative offices near the front lines with units on both sides of wars. The function of these bodies would be to serve as neutral collection and screening points for charges of war crimes, supplementary investigative units for gathering evidence, and, most importantly, communications centers for relaying information about charges, progress of trials, and transmission of evidence from one side to the other. There will be practical problems in functioning as alien forces among military units, but they will not be more difficult than the problems of the International Red Cross and members of the press, all of whom function adequately under such circumstances. Only in cases of gross negligence by the military powers themselves would there be any justification for, or need for, war crimes trials by international bodies following the war.

Dealing adequately and fairly with the crimes of wrongful war and war crimes is certainly difficult and complex. There is little doubt that the efforts of the past are seriously flawed and inadequate. It would take a great deal of both time and effort to encourage a workable system to evolve. It will take many years and a good deal of earnest discussion and skilful pleading. But, given the seriousness of the issues involved, and the consequences for human decency, there is every reason to undertake such an effort.

CHAPTER 13
Demobilization

I Veterans and non-veterans

Those who are able to serve in the military during wartime and are called upon to do so can be divided in two: viz., those who serve and those who do not. For both these groups, the effects of the war upon their lives is likely to continue long after the conflict ends. These effects are not confined to them, but ripple through the larger society. Those whose lives are twisted by war will change the lives of those around them. Those who have become fugitives through evasion of war service will form an underculture, and create an unseen and alien, yet felt, presence. Different nations respond to these groups in different ways. The United States has, for example, been quite generous to both. The government has elaborate systems of benefits for veterans. These are so elaborate that the US Veterans' Administration is one of the largest and most expensive branches of government. It is commonly known that the costs of veterans' benefits following American wars exceed the costs of the war themselves and continue for astonishingly long periods of time.[1] In a different way, the United States is generous to those who refuse to serve as well. Its usual practice since the time of George Washington has been to declare some form of amnesty following wars and rebellions.[2] Even those cases that come to trial are rarely punished to the full extent the law allows. Part of the desire to return to normalcy following wars is to forgive and forget those who have been at odds with the nation's aims during wartime.

Common sense is wedded to the view that something is owed to returning veterans but it is unclear exactly what this is or what the basis of such obligations are. In contrast, ordinary opinion is more unsettled over the question of those who did not serve. It seems unfair, on the one hand, to those who sacrifice to allow those who did not to escape scot free. On the other hand, those who refuse to serve sometimes have compelling reasons for their acts. To try to resolve these issues, it is well to begin by examining the kinds of sacrifices veterans have made and the sorts of problems they are likely to encounter on return to civilian life.

II Experience of veterans

Rhetoric about the sacrifices and suffering of veterans is all too common. Yet part of what makes dealing with returning veterans so difficult is the quite diverse range of experiences they have had. Some will have spent their military service in harsh and inhospitable surroundings experiencing little but pain and fear. Many others, however, will find themselves in more comfortable or even pleasant surroundings with little to do and many opportunities for adventures undreamed of at home. In modern, technologically advanced armies, relatively few military people are directly exposed to the rigors of combat. In the NATO forces, for example, this works out to some three support personnel for each individual engaged in fighting.[3] This ratio will doubtlessly differ for other armies. The fact remains that the majority of those who enter the military spend their time in relative comfort, often accumulating useful skills along the way.

Most veterans have benefited in some fashion or other from their experiences, and many have likely enjoyed a net gain as the result of their military service. It is rare to find a veteran who actually admits to have regretted his experience. Even those who underwent a good deal of hardship believe their experience was valuable – even precious – though few would wish to repeat it. The values likely to be gained are quite varied but include those of adventure and camaraderie, the acquisition of useful skills, honor and prestige at home, increased maturity and self-knowledge – and the not inconsiderable material benefits that many nations bestow on those who have served. Few veterans do not enjoy at least some of these gains. We should neither overstate the sacrifices required of those who served in the military nor understate the considerable benefits that many derive.

Still, almost all veterans have paid some price for their service, however easy their circumstances in the military. Many lacked any overpowering wish to serve in the military and would not have served if it could have been easily avoided. Almost all will have had their normal lives interrupted and have lost some portion of their life that they would have preferred to have devoted to other things. They certainly suffered from separation from their family, friends, and familiar surroundings. In addition, all found some aspects of military life itself unpleasant. All suffered from fear and uncertainty about what would happen to them in the military, and many underwent the risk of injury and death. Of course, many were injured, both physically and psychologically. The dangers and stresses of military life are considerable and extend far beyond the risks of combat. The military makes use of a great deal of dangerous equipment, often in adverse circumstances. The risks of accidents and mishaps are great and account for large numbers of those incapacitated as the result of military life. There are, in addition, the risks associated with boredom and dislocation – and the effort to find release from these in drugs or alcohol. There have probably been nearly as many lives destroyed as the result of immersion in these substances as on the battlefield itself.[4] These hazards are little recognized and usually understated, but account for substantial amounts of the human misery that results from military service.

The diverse experience of veterans and the complex and varied nature of the problems they may bring back with them present unusual challenges for their home society. Veterans will present difficulties that cannot be ignored, yet it is difficult to know how to respond to them.

III Duties of society to veterans

It is commonplace to observe that those who serve have a duty to do so. It is equally common to observe that society nonetheless has duties to them in turn. What is not common is a clear understanding of the basis and nature of these reciprocal duties.

If a nation as a whole is under threat, the citizens within it will be threatened in their lives and well-being. In the event of invasion and conquest all will suffer, though not all to the same degree. Further, those who have prospered under the status quo will have more to lose than others. The rich may have more to lose than the poor, but the poor are more likely to experience intense personal hardship. Differences cut across other dimensions as well. The very old, the infirm, and the very young are most likely to suffer in the event of invasion and conquest, but the vigorous and youthful are the ones to be called upon to resist attack. Given the nature of things, it will be nearly impossible to apportion suffering, interest and ability to serve so that these all balance out equally. The central fact is that all have an interest in resisting invasion and conquest, but each has differing abilities to actually fight for the cause. It would be unfair and unjust to call upon the young and vigorous to serve if all citizens are equally capable of doing so. Since capabilities vary, and since fighting in self-defense is justified, it becomes morally permissible to call upon some, those who are able, to carry the burden of fighting in self-defense. It would not be if it were feasible to arrange matters in some other way. This situation is made more morally palatable by the fact that this distribution is largely a matter of chance. Those who are now young could have been old and infirm – or could have been a few years younger – when the war occurred. Or, they might have suffered some illness or misfortune that could have prevented them from going to war. So society is not unjustly and unduly singling them out for special hardship. Since it is necessary to defend the nation, and since they are able to do so, they have the duty to do so.[5]

Nontheless, because the nation has benefited immeasurably from the service of its soldiers, who have made sacrifices to serve, it has duties to its military as well. But what is the nature and extent of its duties toward veterans? Should the nation pamper them for the rest of their lives, make them rich, provide recompense for their suffering, try to restore them as near to normal functioning as possible, or simply offer them gratitude and parades?

We can gain clarity on this issue by returning to the old standard lifeboat example. A number of people are on a lifeboat some miles from shore, with little hope of being found and no oars or sail. The only way to get ashore is for someone to go over the side to push the boat in. This will be arduous because of the distance and the cold water, and risky because of sharks in the area and the danger of being gashed by submerged rocks. Of the survivors, some are injured and others are elderly. Only one is fit and capable of the strenuous pushing. The hearty individual goes overboard and gets all to safety though he does suffer serious gashes from the rocks and severe fatigue from his task. He clearly had a duty to take on his chore because of the need to help others, though he also saved himself. The others clearly had the right to expect his help, but what do they owe him in return?

It does not seem that they are obligated to make him rich or become his servants for life. Neither do they owe him restitution. It is true that more was demanded of him than of the rest, but there is nothing wrongful or unjust in the sacrifice required of him. There is no unjust loss on his part for which he can claim damages. He did nothing more than what was legitimately required of him.

Still, are those saved obligated to provide him with more than gratitude? Should they provide him with care for the injuries he received? Although his sacrifice was just, he made a sacrifice which was greater than that of anyone else, and it was for their sake. Also, although it was his duty to take on his burden, he did not deserve it in the sense that he had done something special (e.g., made a promise or had done something wrong) to bring this duty about. It was just a matter of chance that he was suited to make his effort. So, his fellow survivors are not entitled to claim that, since he was only doing what was required, he deserves nothing from them. They are at least obligated to provide him with care for his injuries.

But what level of care is required? What should the goal of this care be? He may claim that he is entitled to be returned to the condition he was in before his venture. This is impossible. He would have suffered anyway whether he had made his sacrifice or not, so they are not responsible for the full extent of his harm. For the same reasons, he is not entitled to be made better than he was before his venture. He is not entitled to treatment for his cataracts while his leg injuries are being attended to. Neither is he entitled to training that will suit him for a better career than the one he pursued before the misfortune. Neither, though, is he entitled only to simple treatment necessary to save his life. Basic care may prevent infection and so save his life, but further operations may be necessary to remove scar tissue so that he recovers the full use of his leg. Basic care is cheap, but reconstructive surgery quite elaborate and expensive. The difference is between the care required to keep him alive and treatment that will allow him to lead as near to a full, normal life as is possible. What he would lose without full treatment is the ability to lead a normal life. There is no need for him to sacrifice this. He is not obligated to give it up or lose it for the sake of others. So they have the duty to return it as far as they are able. It is impossible to return the time lost or the anguish suffered, but, if it were possible to return these, it would be their duty to do so.

In the past, with veterans, there has been a tendency to focus only on physical loss and on physical reconstruction. In recent years, however, authorities have become aware that the psychological losses of veterans can be as great and as devastating as the physical. The trauma and anguish of wars, not just the battles but the inhospitable conditions and continual fear of death or dismemberment, can generate psychoses that may remain with the individual for years. There are other causes of psychological difficulty as well. Wars offer inexperienced individuals both the opportunity and the motive to become deeply involved with drugs and alcohol. The numbers involved here are not inconsiderable. Another kind of stress can only be called moral. It is not easy for most to kill or maim other people, and many experience great anguish for having done so, even 'in the line of duty.' Or, in war there are countless opportunities for mistakes to be made that result in death and destruction, sometimes to one's own people. Often the moral standards that an individual was bred to respect are set aside during warfare so that he experiences anguish when he returns to his home environment and tries to re-adopt the values that he once held.

Because awareness of these problems is recent and the development of instruments to measure them is more recent still, it is difficult to find data on these matters for most nations and for most wars. However, experts estimate that as many as a third of the American Vietnam veterans suffered psychological difficulties sufficient to warrant professional treatment.[6] Signs of these strains included significantly higher rates of arrest, alcoholism, automobile accidents, and unemployment among returned veterans than their peer group as a whole.

Even those who were not gravely disabled have had significant difficulty readjusting to civilian life. Time in the military, particularly during warfare, is often totally unrelated to the individual's civilian life before or after. It is like a slice taken out of his life. The individual is removed from family, friends, home environment, educational and career opportunities and given an essentially new identity living under rules quite different from those of civilian life. He lives in an entirely alien world, particularly if involved in war, where killing and death are expected and accepted, and life is lived on the most primitive level. Yet this world is captivating because it is so vivid and elemental and because strong, automatic reflexes must be developed in order to survive.[7] The individual becomes used to the danger and the rushes of adrenalin. This intense existence exerts a strong pull, and minor incidents in later life can often throw veterans back into it. This period of exchanging one world for another is difficult and uncomfortable for many. Individuals often feel permanently altered by their experiences and unable to return fully to their old, normal patterns of life.

These problems stand in the way of normal living in as substantial a fashion as does physical impairment. We have the obligation, then, not only to understand these maladies better but also to develop more effective and more systematic means of dealing with them. We leave it to experts to decide exactly what measures should be taken. It is clear, though, that the military should screen veterans about to be discharged for psychological difficulties as well as drug dependency; and that more active measures need to be taken to ease the transition back to civilian life. Laws mandating that those who left jobs to serve have first priority in receiving them back are a step in the right direction, but it would be useful for the military to establish community educational and job placement centers as well. These measures would be costly, of course, but the expense is insignificant in relation to other military costs and in relation to the costs of treating the physical illnesses of veterans. The cost would also be less than the social costs in terms of crime, accidents, and addiction, that otherwise would be incurred in the absence of these measures.

But counselling by itself is unlikely to be able to meet the full emotional and social needs of veterans. Psychological counsellors are not available for providing friendship or a supportive social environment. These can only be provided by private individuals acting on their own initiative. In the past, large families and stable communities were available to provide these kinds of support. Today, with the greater mobility and social fragmentation of modern societies, this kind of support is no longer available. There are two important resources to draw upon, however. The many groups that form spontaneously to send food, bars of soap, and letters to soldiers overseas could also be tapped to provide them with support when they return home. The same initiative and concern could easily be rechanneled for these other purposes. Another valuable resource is other veterans, those who returned earlier and have been through the period of readjustment. The model here would be something like veterans' clubs that often sprout up after wars, but with a social rather than the usual military and chauvinistic focus.

Lastly, if we may be permitted to say so, adoption and enforcement of rigorous moral standards by military organizations will play its role in forestalling or alleviating some of these problems. Those who can feel that they have conducted themselves honorably, with due concern and respect for others, are less likely to engage in the kind of activity that will lead to guilt and trauma later on. Also, and importantly, they will have a clear conscience about their activities during war and a better image of themselves and their role. This applies not only on the personal level but on the institutional level. Veterans can take comfort in having served with organizations that did not waste human life carelessly or show lack of respect for others. There will be the inevitable tragedies and accidents, of course. No one can expect that veterans will return completely free of trauma, but military organizations that follow moral precepts in the conduct of their operations can expect to mitigate the later suffering of those who are required to carry them out. On a more elevated level, nations can assist with the problem by taking care to get involved only in wars that are clearly morally justified. For example, a substantial portion of the difficulties of American Vietnam-era veterans was due to the fact that they were the instruments of a war that was suspect both in its goals and in the manner in which it was executed. Nations owe it to their soldiers, in a variety of ways, to use them only in wars with clear justification.

IV Those who do not serve

Wars generate varying numbers of those who refuse to serve. Sometimes they can form a substantial underclass living as fugitives either at home or in exile abroad. Part of the business of finishing wars includes determining how to deal with these individuals. The issue is a complex and difficult one not only because the motives and circumstances of those who do not serve will differ greatly from individual to individual, but because differing wars will be more or less justified and more or less popular. We have two issues to confront. The first is whether it is permissible to grant general amnesty after wars are finished to those who failed to serve. The second is the problem of how to handle individuals on a case-by-case basis in situations where general amnesty is not granted.

It is not uncommon for nations to grant general amnesty to draft-dodgers and deserters following wars. A number of factors explain the popularity of this practice. One is simply a desire to return to normalcy as quickly as possible once fighting has ended and to attempt to erase sources of bitter memories or divisiveness. A substantial subclass of fugitives generated by war would be a perennial source of social turmoil and could easily continue to serve as a reminder of the conflict's bitterness. Social repercussions will be magnified because the ranks of this group will be filled by people whom society does not commonly think of as criminals. Rather, it will include large numbers of people with strong family ties whose members will not merely be among the social underclasses, but will range also into the middle and upper classes of society. These will include people with contacts, education, and a well-developed sense of self-respect. Such individuals will not necessarily remain sullen and quiescent or keep their antagonisms to themselves. Erasing this class by granting it amnesty would not only remove a source of tension, but serve as another prominent sign, along with discharged veterans and retooled manufacturing plants, of the war's end. Society will also find it difficult to think of these individuals as criminals because there will be no easily identifiable person whom they have clearly harmed. Criminals of the usual sort normally wrong someone or a few specifiable individuals, and the harm that is done to their victims is readily apparent to all. Those who wrongfully resist serving, however, harm all the members of society by failing to aid in their defense. In addition, the harm caused by their refusal will not be readily perceptible, as a knife wound or empty pocketbook would be. The harm is so general and diffuse as to be invisible.

In addition, once the emergency of war is over, there will be a less perceived need to continue to enforce the rules about the draft and military service. These laws are usually created to provide response to particular occasions. When these occasions are past, the rules will seem superfluous. Then, too, it will be difficult to see how enforcement of the rules will have any clear deterrent effect. Since there will be no foreseeable occasions when there will be a need to enforce the rules, there will be little need to deter individuals from breaking them.

A final consideration arises from the legitimately controversial nature of war. Citizens will often disagree about wars and some will refuse to serve on grounds that a particular war is wrong. Even justified wars may well be unpopular if the connection between the threat they encounter and the lives and security of individual citizens is remote. Amnesty following a war that is unpopular and controversial could well be a gesture of respect and reconciliation for those who have honestly disagreed. In democracies, where citizens are encouraged to form views and act upon them, it can be expected that there will be disagreement and often strong reasons of conscience for not simply following the will of the majority.

These reasons and motives are countered by a number of persuasive arguments supporting punishment, even after the war is clearly ended. The first argument is based on the idea that, in cases of justified war, the community has a right to expect military service from those capable of it. Those who fail to serve have violated this right, a right connected to the most serious matters of life. A natural argument would claim that those who possess a basic right of this sort have the right also, and even the duty, to enforce that right by punishing offenders.

A second argument is that those who do not serve are being unfair to those who did and who suffered because of their service. Those who have been singled out will feel unjustly treated if others escape similar duties without suffering any ill repercussions.

Lastly, it could be argued that a failure to enforce laws, particularly under conditions of great social concern and visibility, will lead to a general breakdown of respect for legal authority. Societies might then have difficulty enforcing laws, especially those concerned with the draft and military service. If individuals have good reason to believe that they will eventually receive amnesty by avoiding punishment until the completion of a war, they are much more likely to do so. The threat of punishment will then lose much of its deterrent effect.

These arguments suggest that the decision to grant amnesty or not must result from a balancing of the reasons on both sides of this issue. The considerations that will help weigh the balance in one direction or the other will be of two sorts. The first will be concerned with how just the war is. Some wars will be quite controversial. Others will clearly be justified, while others will clearly not be. In cases where wars are clearly justified, the balance will weigh considerations of justice against those of benevolence or goodwill. The second sort will be concerned with the urgency of the war. As the situation becomes more urgent, considerations of justice will weigh in the balance more heavily. That is, where the threat to life and well-being is immediate and quite serious, it will be less easy to justify benevolence towards those who failed to meet the threat than in cases where the threat was minor though real.

Questions of the extent of disobedience will also be relevant. It will be more difficult to ignore very widespread and virulent disobedience than a few isolated cases, simply because the challenge to the authority of law will be greater and because the larger numbers will cause greater awareness of the violations.

Even in just wars, however, the need to return to normalcy as quickly as possible and the importance of uniting citizens to build a credible peace will have some weight. Any war disrupts the social and economic life of a nation to some extent. Large numbers of individuals in their prime of life will be removed to engage in military service. This will place stress both on the economy and on the larger social fabric. Wars also demand that substantial segments of the economy be rechannelled to serve the war effort. Finally, wars often result in much physical damage to a nation's people, its resources and its economic infrastructure. The economic and social stresses are likely to result in political and personal stresses as well, particularly where wars are controversial, and most particularly where the nation is on the losing side. Following such a war, it may be especially important to rebuild the country for conditions of peace. Amnesty for those who failed to serve may well further the efforts of returning to normalcy by restoring the individuals and their families to full, loyal members of the community, making their talents and energies available for the restoration effort. It would also spare the community the complication and bitterness likely to result from efforts to apprehend these individuals and place them on trial.

There is no exact formula for deciding how this balancing should be done. An important factor to keep in mind is that the choice to be made is not always the simple, stark one between full punishment and no response at all. A large number of intermediate options are available, including trial and pardon, community service, or symbolic reaffirmation of loyalty to the community through, for example, the process of reapplying for citizenship. These intermediate responses will often be attractive because they will strike a finer balance between the demands of justice and the desire for benevolence and mercy. They have value also in requiring something from the individuals themselves. The gesture of reconciliation should not be one-sided but should include a positive response from both sides.

However, what seems more important than the actual community response is the way in which the decision is made. Announcing a national policy of granting amnesty following wars, or even establishing a clear pattern or tradition of doing so, is likely to be counterproductive simply because it will undermine the deterrent effect of punishment. More important, though, since the community as a whole has been wronged, it is appropriate that the community should collectively decide whether to grant amnesty or not. The directive to do so should not come from on high but should be the decision of ordinary citizens by means of a plebiscite or referendum. That way, a gesture of reconciliation would genuinely come from the entire community, and a decision to withhold amnesty should represent a common decision that justice should outweigh benevolence.

In the aftermath of some wars, a policy of wholesale amnesty will be ill-advised. In these instances, it will be preferable to make decisions on a case-by-case basis. This approach will be preferable particularly when certain conditions obtain. Most cases, for example, will not clearly pit justice against benevolence. For one thing, wars will not often be obviously and incontrovertibly justified. There will be relatively few instances where the nation is responding to a clear act of aggression. Most often, the war will be fought in the name of security, a response to a threat that may be more or less remote and may or may not form a link in a chain of threats which could eventually decisively undermine security. A revolution in a small and distant country, for example, may cause it to ally itself with one's enemy and this alliance may in turn form part of a chain of events that will lead to decisive insecurity. Is war against this nation, aimed at breaking the causal chain, legitimate? Or, are fears about events in a distant country simply the expressions of arrogance and paranoia? There is clearly room for honest disagreement on such issues and, just as clearly, room for the principled conviction that war in this case would be morally wrong and should not be supported.

An additional cause of controversy and unease is that the motives and circumstances of those who do not serve will be quite varied. Some will refuse to serve from firm moral conviction resulting from religious beliefs, others from a general philosophical commitment to pacifism, and others from the simple conviction that the war in question is morally unjustified. Some will avoid service from concern for the welfare of those dependent on them. Yet others will refuse for less lofty reasons, some from simple cowardliness or selfishness.

It is difficult to know what to do about any of these cases in part because it will often be difficult to separate selfish individuals from those whose refusal results from honorable motives. The rhetoric of moral purpose will easily be available to all. Balanced against this is

the respect that should be accorded the individual and the importance of his being able to act in accordance with his conscience. An
individual who is forced to act in ways contrary to his moral values is
harmed as surely as if he were injured physically. Recognizing this,
the community needs to deliberate whether, and under what circumstances, it is justified in imposing this harm on a given individual.
In addition, as Mill argued forcefully, sometimes the individual may
be correct in his views and the community wrong. In such cases, it is
important that his arguments be heard. Even if he is mistaken, it is
still important that he have his say so that all options are examined.
Most frequently, though, as in the case of wars fought for reasons of
national security, there will be no obvious right or wrong and no way
to clearly resolve the issue in favor of one side or the other.

The problem is that nations must finally make decisions and act
upon them. They must decide whether to go to war or not. They
cannot expect that all citizens will agree with the final decision and
must expect that there will be some who have principled objections to
any course of war. Thus, nations will need to decide to what extent
they can and ought to accommodate such dissent.

In some ways those who refuse to serve for reasons of conscience
will be little different from those who are physically handicapped.
Both groups are in some sense unable and unfit to serve. One thing to
notice is that the meaning of 'unfit' is quite elastic and expands or
contracts in accordance with circumstance and need. Physical standards tend to become quite high in times of peace or low-intensity
war and thus the pool of unfit becomes quite large. In times of great
need and pressing emergency the standards are, often informally,
lowered substantially. Larger numbers of people are conscripted,
and the pool of unfit contracts.[8] Nations are thus able to exercise
discretion in determining what groups of people are unfit to serve
and often do so. In the last days of the struggle of Nazi Germany
during World War II, almost all were deemed fit to serve, and the
pool of unfit was small indeed. In many cases, the harm to personal
integrity caused to an individual by forcing him to act against his
conscience will be greater than that suffered by the person whose
incapacity is physical. For this reason, claims of moral harm should
sometimes take precedence over those of physical incapacity. This is
particularly so since many of those with physical defects are eager to
be allowed to do whatever they can to help the community. They
often have strong desires to be useful and productive. It is also the

case that military services can make use of many physically handicapped with only minimal readjustment, particularly since modern technology often eases the physical demands on soldiers. Hence there is little reason why handicaps of conscience should not be given greater accommodation than physical handicaps. Of those with handicaps of conscience, fairness would require that individuals handicapped by religious or ideological aversion to all war should be given greater accommodation than those who have qualms about the justification of particular wars. This is because the members of the latter group will have to acknowledge that they may possibly be mistaken in their views about the wars they are questioning. Further, this group is less likely to be seriously harmed by being required to serve than is the former group.

For clarity's sake we should note that those who object to particular wars may object either because the war itself is wrong and should not have been fought or they may object to the way in which it is being fought. Only the former group has a clear ground on which to avoid service altogether. Members of the latter group can continue to carry out their protests against unjust ways of fighting wars while in the military and, in fact, may have greater success in pressing their case once in the service. In any event, in the extreme situation, they can always refuse to obey immoral commands or refuse to take part in immoral ways of fighting.

Even in cases where a nation cannot accommodate individuals to the extent of allowing them to avoid military service altogether, they can be accommodated in other ways. They can be allowed to refuse to carry weapons, for example, or to take part only in humanitarian activities, such as providing medical care. More importantly, however, each individual should have the opportunity to make his case before the community that is requiring service of him. One way of doing this would be to have hearings, modelled on the pattern of jury trials, that would be open to the public and news media and which would be decided by juries drawn anew for each case, using the same system as is in place for courts of law. Counsel would be available for those who are ill-equipped to present their own views in persuasive fashion. Results of these hearings could then form a body of case law that would build and change as individual cases were heard. This system would allow each person the occasion to state his position as cogently as he can before the community as a whole, and allow the community sufficient flexibility to respond to each case on its merits and not be bound by restrictive regulations.

Individuals presenting their cases in this fashion should recognize that the community is not bound to accommodate their views, however sincerely they hold them. Indeed, sincerity is not always the main issue, though it may be paramount for the individual himself. The community need not accept reasons which are persuasive and compelling for a given individual. Thus, they need not accommodate cases that are irrational, bizarre, or based on alien values. There are limits to the degree of toleration which any individual can require of others. In extreme cases, giving the individual a fair and thorough hearing may be all that a given community can or should be able to do.

Both those who serve and those who do not serve provide complex and difficult problems for their nations. In this chapter we have tried to show what these issues are and to suggest means for handling them. What is most important, however, is to acknowledge that these two groups will remain after wars are finished. They are part of the legacy of all wars, and moral sensitivity, as well as community prudence, requires that they not be ignored.

Conclusion

Military Ethics is not something new on the scene. Soldiers from antiquity have embraced codes and aspired to ideal standards of behavior. Indeed, it has appeared to most that it is just because the circumstances of war are so tumultuous, and its demands on human endurance so great, that strict integrity of thought and action must be required of those who fight. Far from viewing the rigors of warfare as excusing barbaric and undisciplined behavior, commentators have seen it as making the highest standards of conduct necessary. There is also precedent for hewing to strict standards of when going to war is justified. Few commentators have agreed with the view that wars are like natural disasters, floods or hurricanes, whose cause and control are beyond the capacity of human management. Both Plato and Aristotle argued that the virtuous state would only go to war for just cause. The tradition of natural law theorists, best worked out by Grotius, has insisted on establishing principles to guide the initiation of war. And, there is evidence that the states of Europe have been influenced in their conduct by these principles – albeit imperfectly.

History alone is thus sufficient to undermine those, the realists and the pacifists, who argue that warfare cannot be controlled by moral precept. They do have some history on their side, however. The wars of the latter part of the nineteenth century and for much of the twentieth have been notable for their lack of moral constraint. What the realists and pacifists fail to understand is that this lack is due to special developments, and that it is possible to come to terms with these developments in a clear-headed fashion and to reintroduce moral constraint. Those who become overly idealistic in their enthusiasm for reform fail, on their part, to understand that these same developments also place limitations on what can be demanded by a military ethics at the present time and fail to understand, as well, that more rigorous standards of conduct will require new social institutions and different political circumstances to support them.

Perhaps the major factor in the recent demise of standards of military ethics in Western Europe at least, is the extinction of small, professional armies and their replacement by huge masses of conscripts. The professional armies of Europe a few centuries ago were almost fastidious in their tender regard for one another. Small groups of men adopting the profession of arms for a lifetime could be expected to develop rigorous standards of personal conduct and could rely on their counterparts in opposing camps to do the same. Indeed, soldiers, or at least officers, of differing armies often came from the same social classes, frequently knew one another personally, and commonly were connected by family ties. Such conditions almost insure the vigorous flourishing of standards of decency. The temporary, impersonal, and heterogeneous nature of mass conscript armies, just as surely, led to a breakdown of these constraints.

Along with these changes came the breakdown of stable groupings of states with shared cultural backgrounds and common standards of international behavior. Fixed groups of European states or Arab states battling with one another could expect their antagonists to share the same traditions of the proper conduct of war. This heritage placed constraints on what could and could not be done in good faith and with the good opinion of one's peers. With the advent of world wars that spread all over the globe and with social revolutions that caused the demise of the ties and classes that supported these traditions, the standards were forgotten.

Technology has also done its part to undermine standards of moral conduct. Weapons that can kill at long range, and indiscriminately, circumvent the inhibitions that can serve to mitigate slaughter when killing must be done face to face. The most mild-mannered bombardier is capable of destroying many more people than the most rampant of Mongol hordes. Slaughter thus becomes easy, both emotionally and physically, and this ease readily short-circuits moral constraint. These weapons, along with the sheer mass of contemporary armies, make it difficult to restrict the area of violence and destruction. Wars of the seventeenth century could be fought on deserted meadows and non-combatants easily spared. Wars of the twentieth century cannot be contained so easily.

In addition, certain types of war have become common and have done their part to destroy moral constraints. Guerrilla wars and civil wars, for example, obliterate the distinction between combatant and non-combatant almost beyond repair. Some are even fought in population centers rather than in empty places. Their very nature causes erosion of the social ties and constraints which ordinarily prevent violence from getting out of hand. In addition to these are the return of what Richard Brandt calls 'serious wars', the wars where nations are fighting for their very existence. Conditions of twentieth century warfare and twentieth century ideologies have caused a return to something approaching the era when nations losing wars were likely to have their inhabitants put to the sword en masse. The citizens of Great Britain were fighting for their lives against Hitler in a manner none too different from the way the citizens of Troy fought against the Greeks. In such wars, where everything is at stake, constraint is likely to be tossed to the winds.

These factors, in addition to hastening the decline of moral constraint, also indicate the conditions and challenges that must be met by any attempt to revive standards of military ethics. Other conditions that must be met derive from the complex demands that modern warfare places on entire nations. The preparation for and conduct of war involves all segments of a nation. Because of this, an adequate military ethics must take account of the role of all sectors of society in military activity. In addition, these preparations must begin long before wars begin and will have repercussions long after they end. Military ethics thus cannot be an ethics only for or of the military, and it cannot be an ethics only of war. It must be an ethics that accounts for all segments of society and their role in military affairs, and it must contain a much larger perspective than only that of war itself.

We believe that the only approach to ethics capable of meeting these conditions is one that takes explicit account of the interplay of the capacities of solitary individuals, the structure of the social institutions in which they act, and the nature of the political and social conditions which enfold them. It is futile, for example, to make demands on individual behavior without taking account of the social institutions that structure this behavior or the social conditions that color it. It is important to understand that this interplay not only places limits on what can be required of people, but it also creates opportunities for change. Understanding what institutions would be required justly and meaningfully to conduct trials for war crimes, for example, not only indicates the limitations on what we can legitimately expect at the present time, under present conditions, but points the direction we must take if we wish honestly to enforce standards of conduct during war.

Our efforts to comprehend the interplay of these factors as they are related to the military have led to the approach based on a full analysis of the conditions under which individuals must act. We believe that this approach not only gives us the tools necessary to understand the present and give direction to future efforts, but it also gives us means of accommodating changes in international conditions as they occur and revising moral conditions to follow suit. It is our hope that such flexibility can prevent moral constraint from getting left behind in the face of future development, as it got left behind by the changes of the nineteenth and twentieth centuries. Our recognition of this complexity and the need for flexibility is in part what prompted us to adopt an essentially utilitarian normative perspective. Rights-based or duty-based moral systems simply do not contain the means of accommodating future change or unexpected complication. They are doomed, in their striving for timeless standards of conduct that can never be breached, to be rooted in a single perspective that is all too likely to be swept away by the tide of events. They require being fixed in particular social institutions that may prove inadequate to new demands and cannot point the way to develop new ones. Finally, by their nature, they demand either full compliance or none at all. Rights, for example, make their claims absolutely. They cannot allow compromise, and they cannot await the development of social institutions capable of supporting them. Only utilitarianism provides this degree of accommodation.

We agree, then, with the ancients that war is too destructive and too demanding to fail to devise moral standards to constrain it. Indeed, these features are accentuated by contemporary warfare. We agree also that individuals are capable of upholding these standards and of preserving their humanity under even the most strained conditions. We insist only that these human capacities must be nurtured and supported by the proper social institutions developed in the right conditions. Failure to understand this will lead either to the despair of those who believe no moral standards can be effective in wartime or to the euphoria of those who would demand too much.

Notes

Introduction

1 One hundred and fifty-eight of the countries of the world are considered to have significant military forces of one sort or another. *The Military Balance*, London, International Institute for Strategic Studies, 1980, p.2.

2 R.M. Hare, *Moral Thinking*, Oxford, Clarendon Press, 1981.

3 Ibid., pp.29–30.

4 Recent and useful discussions of the realist theory include: Charles R. Beitz, *Political Theory and International Relations*, Princeton, Princeton University Press, 1979, pp.15–27; Stanley Hoffmann, *Duties Beyond Borders*, Syracuse, NY, Syracuse University Press, 1981, pp.45–55; and Michael Walzer, *Just and Unjust Wars*, New York, Basic Books, 1977, pp.3–20.

5 See Beitz, op. cit., pp.20–5.

6 A particularly thorough and sensitive discussion of pacifism is developed by Cheney C. Ryan, 'Self-defense, pacifism, and the possibility of killing,' *Ethics*, vol.93, 1983, pp.508–24. Also useful are: Jan Narveson, 'Pacifism: a philosophical analysis,' *Ethics*, vol.75, 1965, pp.259–71; Tom Regan, 'A defense of pacifism,' *Canadian Journal of Philosophy*, vol.2, 1972, pp.73–86; and Judith Jarvis Thomson, 'Self-defense and rights,' *The Lindley Lectures*, Lawrence, KA, University of Kansas Press, 1976.

7 See Walzer, op. cit., pp.127–222.

8 Recently, there has been some interest in yet another approach, the attempt to found morality on an appeal to virtue. The most vigorous proponent of this movement is Alasdair MacIntyre in his felicitously titled *After Virtue*, Notre Dame, University of Notre Dame Press, 1981. There are several strengths in such an attempt, but also a flaw which, for our purposes, is fatal. MacIntyre himself acknowledges that a virtue-based morality can only be instituted in a small, closely knit, homogeneous community, one in which there is a core of tightly held values. Such conditions are nowhere near obtaining in most areas of contemporary life in which there are serious moral problems. A virtue-based military morality might have been feasible in the era in which armies were small groups of professionals. Contemporary mass armies composed of large numbers of individuals, serving for brief

periods of time, drawn from diverse walks of life are unlikely institutions for developing the sort of community MacIntyre requires. In addition, the moral decisions of individual soldiers in stressful situations are only one arena of military ethics. Decisions about strategy, armaments, whether or not to go to war, are all morally sensitive, and it is difficult to understand how a virtue-based morality could be equipped to deal with them.

MacIntyre does have a point worth taking seriously. It is that it is not sufficient to list a set of rules of correct conduct. It is also essential to pay attention to the training, habits, and character of those whose moral conduct we wish to guide. This, however, is not essentially different from Hare's argument that moral training is a central part of morality. We hope to develop these ideas further and show how they can be applied in a specifically military context.

9 John Rawls, *A Theory of Justice*, Cambridge, MA, Harvard University Press, 1971.

10 Thomas Scanlon, 'Contracturalism and utilitarianism', in Amartya Sen and Bernard Williams (eds), *Utilitarianism and Beyond*, Cambridge, Cambridge University Press, 1982, pp.103–28.

11 See Beitz, op. cit., pp.127–43 and Henry Shue, *Basic Rights*, Princeton, Princeton University Press, 1980, pp.13–34.

12 Alan Gewirth, *Reason and Morality*, Chicago, University of Chicago Press, 1978; David A.J. Richards, 'Rights, resistance, and the demands of self-respect,' *Emory Law Journal*, vol.32, 1983, pp.405–35; Ronald Dworkin, *Taking Rights Seriously*, Cambridge, Harvard University Press, 1977; and Alan Donagan, *The Theory of Morality*, Chicago, University of Chicago Press, 1977.

13 Rawls, op. cit., pp.541–4.

14 Scanlon makes this argument in 'Contracturalism and utilitarianism,' pp.124–8.

15 Rawls, op. cit., p.27 and Richards, op. cit., pp.35–7.

16 Hare, op. cit., pp.25–8.

17 Dworkin, op. cit., pp.235–7.

18 Rawls, op. cit., p.31.

19 Dworkin, op. cit., pp.237–8; Richards, op. cit., pp.413–17; and Amartya Sen and Bernard Williams (eds), op. cit., pp.4–5.

20 Scanlon, op. cit., p.103; Dworkin, op. cit., pp.94–100.

21 Rawls, op. cit., pp.161–75.

22 Dworkin, op. cit., pp.223–39.

23 Ibid., p.234.

24 R.M. Hare, *Freedom and Reason*, Oxford, Oxford University Press, 1965, pp.171–3, but also see pp.117–29 in *Moral Thinking*.

25 Dworkin, op.cit., p.234.

26 Hare, 1981, op. cit., p.105 and Richard B. Brandt, *A Theory of the Good and the Right*, Oxford, Clarendon Press, 1979, pp.70–87.

27 Remember that Rawls, for example, wants to rule out certain preferences and characteristics, op. cit., p.31.

28 Hare, 1981, op. cit., pp.138–9.

29 Ibid., p.123.
30 Ibid., pp.120–9.

CHAPTER 1   *The justification of standing armies*

1 *The Military Balance 1980–1981*, London, International Institute for Strategic Studies, 1980, p.2.
2 Ibid., p.68 (Fiji), p.29 (Luxembourg), p.5 (USA), and pp.61 & 64 (People's Republic of China).
3 *World Armaments and Disarmament: Stockholm International Peace Research Institute Yearbook 1981*, London, Taylor & Francis, 1981, p.xvii and *World Armaments and Disarmament: Stockholm International Peace Research Institute Yearbook 1983*, New York, Taylor & Francis, 1983, pp.129–30.
4 See, for example, Marek Thee, 'Militarism and militarization in contemporary international relations,' in Asbjorn Eide and Marek Thee (eds), *Problems of Contemporary Militarism*, New York, St Martin's Press, 1980, p.18 and Robert Matthews, 'National security: propaganda or legitimate concern,' pp.140–7 in the same volume. In addition, see Ralph E. Lapp, *The Weapons Culture*, New York, W.W. Norton, 1968, pp.174–5.
5 *The Military Balance 1980–1981*, op. cit., p.96.
6 *World Armaments and Disarmament*, pp.xvii & 15.
7 Ibid., p.16.
8 William Gutteridge, *Military Institutions and Power in the New States*, New York, Frederick A. Praeger, Inc., 1965, pp.37–46 and Charles F. Cortese, *Modernization, Threat, and the Power of the Military*, Beverly Hills, CA, Sage Publications, 1976, p.21.
9 See, for example, Gutteridge, op. cit., and Cortese, op. cit. The Cortese volume includes a useful list of references.
10 Cortese, op. cit., p.59.
11 Samuel P. Huntington, *The Common Defense*, New York, Columbia University Press, 1961, pp.443–4.
12 Eide and Thee, op. cit., pp.376–7.
13 A listing of useful articles on pacifism and the justification of self-defense is contained in footnote 6 of the Introduction.
14 See the response to this claim in Cheney C. Ryan, 'Self-defense, pacifism, and the possibility of killing,' *Ethics*, vol.93, 1983, p.510.
15 The argument comes from Jan Narveson, 'Pacifism: a philosophical analysis', *Ethics*, vol.75, 1965, pp.259–71. The response developed here is essentially the same as Ryan, op. cit., p.514.
16 A number of authors who base their findings on appeal to rights arrive at similar conclusions. See Richard Wasserstrom, 'Book Review: Walzer: *Just and Unjust Wars*,' *Harvard Law Review*, vol.92, 1978, pp.538–9, 540–1, and 544; David Luban, 'Just war and human rights,' *Philosophy and Public Affairs*, vol.9, 1980, pp.170, 174–5; and Charles R. Beitz, *Political Theory and International Relations*, Princeton, Princeton University Press, 1979, pp.121–3.
17 An example of this approach can be found in Norman Freund, 'Nonviolent national defense', *Journal of Philosophy*, vol.13, 1982, pp.12–17.

From a different perspective, Douglas Lackey argues that a nation which lacks nuclear weapons is less likely to provoke nuclear attack by those who do, in 'Missiles and morals,' *Philosophy and Public Affairs*, vol.11, 1982, pp.189–231.

18 Carl von Clausewitz, *On War*, ed. & intro. by Anatol Rapoport, Baltimore, MD, Penguin Books, 1972.

19 Even those commentators who argue that most governments do not have the right to be free from intervention agree that the circumstances under which such surgical intervention could be carried out are quite rare. See Gerald Doppelt, 'Statism without foundations', *Philosophy and Public Affairs*, vol.9, 1980, p.400 and Charles R. Beitz, 'Nonintervention and communal integrity,' *Philosophy and Public Affairs*, vol.9, 1980, p.388.

20 This claim is made most vigorously by Walzer, op. cit., pp.52–61.

21 These issues are nicely analyzed by Thomas C. Schelling, *Arms and Influence*, New Haven and London, Yale University Press, 1966.

22 See the comments of Dr Frank Barnaby, director of Sipri, in *World Armaments and Disarmament 1981*, op. cit., p.xvii.

23 This claim is echoed by J.H. Hare and Carey B. Joynt, *Ethics and International Affairs*, New York, St Martin's Press, 1982, p.104.

24 The problems of distinguishing offensive from defensive weaponry are carefully analyzed by Marion William Boggs, *Attempts to Define and Limit 'Aggressive' Armament in Diplomacy and Strategy*, Columbia, MO, University of Missouri Press, 1941.

25 Pitfalls of the search for ever increasing security are nicely explained by Hare and Joynt, op. cit., pp.141–51.

26 The difficulty of 'putting oneself in the place of the other' on the level of international relations, and its moral necessity, is discussed by Hare and Joynt, op. cit., pp.163–83.

CHAPTER 2   *Issues concerning military personnel*

1 Michael D. Bayles, *Professional Ethics*, Belmont, California, Wadsworth 1981, p.7.

2 Ibid., p.8.

3 Morris Janowitz, 'The logic of national service', in Martin Anderson (ed.), *The Military Draft*, Stanford, CA, Hoover Institute Press, 1982, pp.420–1.

4 Melvin R. Laird, 'People, not hardware: the highest defense priority,' *American Enterprise Institute Special Analysis*, Washington, D.C., American Enterprise Institute, pp.1–24.

5 Walter Y. Oi, 'The economic costs of the draft,' in Anderson, op. cit., pp.317–46.

6 Milton Friedman, 'Why not a volunteer army?,' in Anderson, op. cit., pp.621–32.

7 Laird, op. cit., p.4.

8 Charles C. Moskos, 'The all-volunteer force,' in Morris Janowitz and Stephen D. Westbrook (eds), *The Political Education of Soldiers*, Beverly Hills, CA, London and New Delhi, Sage Publications, 1983, p.313.

9 Robert K. Fullinwider, 'The all-volunteer force and racial balance,' in Robert K. Fullinwider (ed.), *Conscripts and Volunteers: Military Requirements, Social Justice, and the All-Volunteer Force*, Totowa, NJ, Rowman & Allanheld, 1983, pp.178–88.
10 Edward M. Kennedy, 'Inequities in the draft,' in Anderson, op. cit., pp.527–9.
11 Ibid., pp.527–9.
12 Moskos, op. cit., p.313.
13 Ibid., p.312.
14 H.B. Simpson, 'Compulsory military service in England,' in Anderson, op. cit., pp.463–77.
15 Laird, op. cit., p.6.
16 D.H. Monro, 'Civil rights and conscription,' in Anderson, op. cit., pp.133–51.
17 Janowitz, op. cit., pp.412–16.
18 Joseph M. Scolnick Jr., 'Case studies – Britain and Canada,' in James C. Miller, III (ed.), *Why The Draft?*, Baltimore, MD, Penguin Books Inc., 1968, pp.91–104.
19 Richard W. Hunter and Gary R. Nelson, 'The all-volunteer force: has it worked, will it work? (Summary),' *Registration and the Draft*, Stanford, CA, Hoover Institute Press, 1982, pp.16–17.
20 George W. Ball, 'The cosmic bluff,' *The New York Review of Books*, July 21, 1983, pp.37–41.
21 Laird, op. cit., p.6.
22 Janowitz, op. cit., pp.420–1. See also Margaret Mead, 'A national service system as a solution to a variety of national problems,' in Anderson, op. cit., pp.431–43.
23 David H. Marlowe, 'The meaning of the force and the structure of the battle: Part I – the AVF and the draft,' in Fullinwider, op. cit., pp.46–57.
24 These passages do not necessarily, and probably do not especially, characterize samurai behavior and ethics before the seventeenth century.
25 Yamamoto Tsunetomo, *Hagakure: The Book of the Samurai*, Tokyo, New York and San Francisco, Kodansha International, 1979, p.17.
26 Ibid., p.20.
27 Richard Storry, *The Way of the Samurai*, New York, G.P. Putnam's Sons, 1978, p.102.
28 Ibid., p.102. Yamamoto Tsunetomo puts it this way: 'It is not good to settle into a set of opinions. It is a mistake to put forth effort and obtain some understanding and then stop at that. At first putting forth great effort to be sure that you have grasped the basics, then practicing so that they may come to fruition is something that will never stop for your whole lifetime. Do not rely on following the degree of understanding that you have discovered, but simply think, 'This is not enough.' One should search throughout his whole life how best to follow the Way. And he should study, setting his mind to work without putting things off. Within this is the Way.' p.31.
29 *The Report of the President's Commission on the All Volunteer Force*, Washington, D.C., Government Printing Office, 1970.

30 Morris Janowitz, *The Professional Soldier*, Glencoe, Illinois, Free Press of Glencoe, 1960, pp.42–3.
31 Ibid., pp.40–1.
32 Ibid., p.41.
33 Ibid., p.50.

CHAPTER 3   *The place of codes of ethics in the military*

1 R.M. Hare, *Moral Thinking*, Oxford, Clarendon Press, 1981.
2 A.J.P. Taylor, *A History of the First World War*, New York, Berkeley Publishing, 1963, p.124.
3 This is not to say that at other times groups of young men will not do positive things.
4 Cleveland W. Feemster (Sgt), 'Ethics and the noncommissioned officer,' *Ethics and the Military Profession*, 1979, pp.2–5. In the same issue see also 'Military professionalism and the emergence of the NCO', by Captain Calvin T. Higgs, Jr., pp.5–18, and 'What is the ethical code of the noncommissioned officer?' by Captain James Narel, pp.19–22.
5 *Prisoner-of-War-Resistance*, Field Manual No 21–78, Washington, D.C., Headquarters, Department of the Army, 1981, pp.5–10.
6 Ibid., p.9.
7 Ibid., pp.9–10.
8 *MQS III Training Support Package: Ethics and Professionalism*, Fort Benjamin Harrison, Indiana, The Military Professional Ethics Division, Training Development Directorate, US Army Support Center, 1983. See also MQS I (1981) used for training ROTC (Reserve Officer Training Corps) students and MQS II (1982) used for training newly commissioned officers.
9 Ibid., Lesson 11, p.10.
10 W. von Bredow, 'The West-German Bundeswehr as an institution of political education', in Morris Janowitz and Jacques Van Doorn (eds), *On Military Ideology*, Rotterdam University Press, 1971, pp.97–115. Von Bredow claims that the German 'moral armament' program became overly anti-communist and tended to explain issues in black and white terms (p.105). In part he attributes the failings of the German program to officers who were (are) not well trained to teach 'civic education'. An earlier and somewhat more optimistic portrait of civic education in the German army is found in Eric Waldman's *The Goose Step is Verboten: The German Army Today*, New York, Free Press of Glencoe, 1964.
11 Ralph Zoll, 'The German armed forces,' in Morris Janowitz and Stephen D. Westbrook (eds), *The Political Education of Soldiers*, Beverly Hills, Sage Publications, 1983, pp.209–48. Like von Bredow, Zoll sees failings in the German civic education process.
        To this must be added that a particular commitment to *politische Bildung* on the part of the company commander would perforce lead to negligence of other areas, given the undebatable fact that he is already overburdened by his wide range of duties. Since a special commitment to *politische Bildung* would in any case turn out negative

in the total evaluation due to the lack of judgmental categories for
this task as well as to the above-mentioned negligence of other tasks,
narrow limits have also been set for any special personal motivation
on the part of individual officers . . . In practice this can be seen in
that (a) political education is cancelled quite frequently, (b) company
commanders transfer the actual instruction to subordinates, (c)
lessons are seldom adequately prepared, and (d) they are frequently
scheduled for 'unfavorable' times. (p.227)

12 Gwyn Harries-Jenkins, 'The British armed forces,' in Janowitz and
Westbrook, op. cit., pp.83–93. Agreeing with this statement, (p.93)
Harries-Jenkins nonetheless notes that '. . . civic education in the British
military has often been based on the premise that personnel, especially
recruits, receive a full education in civilian life' (p.83). He also notes that
the reluctance of the British military to support 'civic education' is due in
part to a confusion between that concept and 'humane education' (i.e.,
education aimed at self-improvement) (p.87). Not being willing to
support the latter non-occupational program, by association, the military
is unwilling to support the former more vocationally oriented program.
Victor Azarya in his 'The Israeli armed forces' (pp.99–127, same
volume) paints a quite different picture of a military organization's
attitude toward civic education.

More than 30 years after independence, despite all the changes
that have occurred in both the armed forces and the parent society,
despite their bureaucratization, their institutional differentiation,
and the formalization of the relations between armed forces and
society, the IDF still acts as a 'school for the nation.' Hence, the
goals of its civic education are different from both its Western and
Soviet counterparts. In Israel, civic education offered by the military
is primarily conceived as a contribution to the entire society; it
maintains the dual 'national service' and 'distinctive superiority'
attributes of the military in society, rather than contributes to
military performance per se. To the extent that military performance
does improve as the result of civic education, this is a welcome side
effect, but it is not the primary objective. (p.100)

CHAPTER 4   *The military and other institutions*

1 Samuel P. Huntington, *The Soldier and the State*, Cambridge, Harvard
University Press, 1957.
2 Ibid., p.83.
3 Ibid., p.83.
4 Ibid., p.83.
5 Ibid., p.84.
6 John H. Garrison, 'Military officers and politics II,' in John F. Reichart
and Steven R. Sturm (eds), *American Defense Policy*, 5th Edition,
Baltimore and London, Johns Hopkins University Press, 1982, pp.760–
7. Garrison holds views similar to our own. Like us, he sharply
distinguishes partisan involvement of the military in political matters

from involvement '. . . in the process of determining how the federal government and American society will allocate values' (p.763). He favors only the latter form of involvement.

7 Morris Janowitz, *The Professional Soldier*, Glencoe, Ill., Free Press, 1960, pp.236–55. In a special report titled 'Can we fight a modern war?' *Newsweek* (July 9, 1984, pp.32–51) interviewed 257 generals and admirals in all four branches of the United States military. The magazine reports: 'Fully 85% of the sample identifies itself on the conservative end of a liberal-to-conservative ideological scale.' As a reflection of this same pattern of preferences, only 4% say they are Democrats while 52% identify themselves as Republicans.

8 Huntington, op. cit., p.72. This is what Huntington calls the representative function of the military man.

9 Jerome Slater, 'Military officers and politics I,' Reichart and Sturm, op. cit., pp.749–56.

10 Huntington, op. cit., p.351.

11 Max Hastings and Simon Jenkins, *The Battle of the Falklands*, London and Sydney, Pan Books, 1983, p.321.

12 The Commission on Freedom of the Press, 'The requirements,' in Bernard Berelson and Morris Janowitz (eds), *Reader in Public Opinion and Communication*, Second Edition, New York, Free Press, 1966, pp.529–34. The Commission says near the beginning of its document: 'Today our society needs, first, a truthful, comprehensive, and intelligent account of the day's events in a context which gives them meaning, second, a forum for the exchange of comment and criticism . . .' In the same volume the Royal Commission on the Press in an article titled 'The standard by which the press should be judged' says: 'The Press may be judged, first, as the chief agency for instructing the public on the main issues of the day. The importance of this function needs no emphasis.' (p.535)

13 Hastings and Jenkins, op. cit., p.376.

14 Frederick Schauer, *Free Speech: A Philosophical Enquiry*, Cambridge, Cambridge University Press, 1982. Schauer quite rightly tells us that some exchanges of communication are counterproductive (e.g., when both sides in a disagreement engage heavily in the use of persuasive language). Nonetheless, he says the following about free speech and its truth gathering tendencies:

> The argument from truth may be based not only on its inherent scepticism about human judgment, but also on a more profound scepticism about the motives and abilities of those to whom we grant political power. The reason for preferring the marketplace of ideas to the selection of truth by government may be less the proven ability of the former that it is the often evidenced inability of the latter.

15 Ronald Dworkin, 'Is the press losing the first amendment?,' *New York Review of Books*, vol.27, no.19, 1980, pp.49–57.

16 Some incidents personally reported to us include those in Vietnam where local commanders were forced to alter their battle plans to protect media

people who had wandered off into dangerous parts of the battle zone. We were also told of an incident (perhaps not uncommon) where the number of media people covering military exercises in Europe was so great that one personnel carrier in ten was used to carry these people and their equipment to the 'battle' zone. That represents a force loss of 10%.

17 Hastings and Jenkins, op. cit., p.374.

18 'Next Grenada may make early editions,' *New York Times*, October 14, 1984.

19 In the US there are such groups as the Rand Corporation, the Hoover Institute, the American Enterprise Institute, to name a few.

20 Paul A.C. Koistinen, *The Military Industrial Complex: A Historical Perspective*, New York, Praeger Publishers, 1980. The complex, Koistinen says, is

> ... made up of a core of about 200 corporations, largely in the heavy and high-technology industrial sector, and the federal executive branch, particularly the president, his chief advisers, and the most important cabinet departments. These two components of the elite have a separate, though hardly unrelated, power base which reflects a very definite middle-to-upper class bias. (p.14)

See also Benjamin Franklin Cooling (ed.), *War, Business and American Society: Historical Perspective on the Military Industrial Complex*, Port Washington, NY, Kennikat Press, 1977. Charles C. Moskos, Jr, 'The military-industrial complex: theoretical antecedents and conceptual contradictions,' in Sam Sarkesian (ed.), *The Military-Industrial Complex: a Reassessment*, Beverly Hills, London, Sage Publications, 1972, pp.3–23. Moskos gives a good overview of various theories concerning the complex.

21 Robert Heilbroner, 'Military America,' *New York Review of Books*, vol.15, no.2, 1970, pp.5–8. J.M. Swomley, *The Military Establishment*, Boston, Beacon Press, 1964. Sidney Lens, *The Military Industrial Complex*, Philadelphia, Pilgrim Press, 1970. Seymour Melman, *Pentagon Capitalism*, New York, McGraw Hill, 1970.

22 Stockholm International Peace Research Institute, *World Armaments and Disarmament, SIPRI Yearbook*, 1983, New York, Taylor & Francis, 1983. For the years 1978–1982 SIPRI puts the US first as an exporter of weapons and the USSR second. The US share of the total (in terms of dollars) is 36.4%. The Russian share is 34.3%.

23 Miroslav Nincic, *The Arms Race: The Political Economy of Military Growth*, New York, Praeger Publications, 1982, pp.111–36.

24 Ibid. In his chapter on the USSR, Nincic says:

> What emerges thus far is a picture of a massive bureaucracy chugging along a steady course that progresses within a relatively narrow band centered on a rising trend. Furthermore, we seem to have come a long way with a single-factor interpretation that would have been possible in the US case. Nevertheless, bureaucratic dynamics should not be endowed with more explanatory power than

the evidence warrants. While Soviet trends do manifest somewhat less volatility than do those of the United States, they are not altogether smooth either. (p.63)

See also Vernon V. Aspaturian, 'The Soviet military-industrial complex: does it exist?', in Steven Rosen (ed.), *Testing the Theory of the Military-Industrial Complex*, Lexington, Mass., D.C., Heath, 1973, pp.103–33.

25 James F. Dunnigan, *How to Make War: A Comprehensive Guide to Modern Warfare*, New York, Quill (William Morrow), 1982, pp.347–74. Dunnigan places US military spending at about 7% of the GNP and the comparable figure for the USSR at 14%. (p.348)

CHAPTER 5 *Just cause of war*

1 Telford Taylor, 'Just and unjust wars', in Malham M. Wakin (ed.), *War, Morality and the Military Profession*, Boulder, CO, Westview Press, 1979, pp.245–58.
2 Malham M. Wakin, 'Introduction to part 2', ibid., p.238.
3 The list of those who criticize this view grows continuously but includes: Michael Walzer, *Just and Unjust Wars*, New York, Basic Books, 1977, pp.85–6; Gerald Doppelt, 'Walzer's theory of morality in international relations,' *Philosophy and Public Affairs*, vol.8, 1978, pp.3–26; Richard Wasserstrom, 'Book review: Walzer's *Just and Unjust Wars*,' *Harvard Law Review*, vol.92, no.2, 1978, pp.536–45; and J.E. Hare and Carey B. Joynt, *Ethics and International Affairs*, New York, St Martin's Press, 1982, pp.55–79. The most striking development of these arguments is found in David Luban, 'Just war and human rights,' *Philosophy and Public Affairs*, vol.9, 1980, pp.150–81.
4 Walzer, op. cit., pp.51–3.
5 Walzer, op. cit., p.113.
6 Luban, op. cit., p.178.
7 Doppelt, op. cit., pp.4–5 and Luban, op. cit., p.169.
8 Charles Beitz, 'Nonintervention and communal integrity,' *Philosophy and Public Affairs*, vol.9, 1980, pp.388–9; Gerald Doppelt, 'Statism without foundation,' *Philosophy and Public Affairs*, vol.9, Summer 1980, p.400 and Gerald Doppelt, op. cit., p.14.
9 See, in particular, Charles Beitz, *Political Theory and International Relations*, Princeton, NJ, Princeton University Press, 1979, pp.65–115. Gerald Doppelt develops a theory compatible with this in 'Walzer's theory of morality,' pp.3–26.
10 *New York Times*, July 11, 1984, p.3, col.4 and October 4, 1984, p.3, col.2.
11 Luban, op. cit., p.171.
12 For a more detailed discussion of this issue see Gerard Elfstrom, 'On dilemmas of intervention,' *Ethics*, vol.93, 1983, pp.709–25.
13 Luban, op. cit., pp.177–8.

CHAPTER 6   *Role of third parties*

1 For information on the varying levels of Cambodian combat with North Vietnam and the Viet Cong, see William Shawcross, *Sideshow*, New York, Simon & Schuster, 1979, pp.96–111 and 128–49.
2 Dale R. Tahtnen notes that Saudi Arabia sent only a small group of reserve forces to the 1973 war in *National Security Challenges to Saudi Arabia*, Washington, D.C., American Enterprise Institute for Public Policy Research, 1979, p.25. It is pointed out in *The Middle East Military Balance, 1983* that the Saudis have limited offensive capacity, would have difficulty in positioning troops against Israel in time of war, and give every indication of preferring to avoid direct conflict. See Mark Heller (ed.), *The Middle East Military Balance, 1983*, Boulder, CO, Westview Press, 1983, p.207.
3 Michael Walzer has a lucid discussion of what we call 'legal neutrality' in *Just and Unjust Wars*, New York, Basic Books, 1977, pp.234–5.
4 Ibid., p.235.
5 Walzer, for example, seems to presuppose this throughout his discussion of neutrality. Ibid., pp.233–50.
6 *Time*, vol.122, no.9, 1983, pp.14–25.
7 See Raymond Bonner, *Weakness and Deceit: U.S. Policy and El Salvador*, New York, Times Books, 1984, or Steffen W. Schmidt, *El Salvador, America's Next Vietnam?* Salisbury, NY, Documentary Publications, 1983.
8 Walzer, op. cit., p.237.
9 Ibid., p.238.
10 Ibid., p.237.
11 Ibid., pp.236–7.
12 Ibid., pp.241, 247–9.
13 Ibid., p.247.
14 Ibid., p.247.

CHAPTER 7   *The enemy*

1 Max Hastings and Simon Jenkins, *The Battle for the Falklands*, London and Sydney, Pan Books, 1983, p.283.
2 John Austin, *How to do Things with Words*, New York, Oxford University Press, 1962, pp.12–24.
3 John Searle, 'A taxonomy of illocutionary acts,' *Expression and Meaning*, Cambridge, Cambridge University Press, 1979, pp.12–20.
4 N. Fotion, 'I'll bet you ten dollars that betting is not a speech act,' in Herman Parret (ed.) *Possibilities and Limitations of Pragmatics*, Philadelphia, John Benjamins North America, 1981, pp.211–23.
5 A non-threatening retaliatory act can be imagined. Just before the peace treaty is signed nation B retaliates by bombing a famous cathedral just to get even for an earlier bombing of one of its famous cathedrals.
6 Michael Walzer, *Just and Unjust Wars*, New York, Basic Books, 1977, pp.208–11.

CHAPTER 8   *Weapons of war*

1 Department of the Army, *The Law of Land Warfare*, FM 27–10, Washington, US Government Printing Office, July, 1956, p.18.
2 Ibid., p.18.
3 Ibid., p.19.
4 Leon Friedman, *The Law of War: A Documentary History*, vol.1, New York, Random House, 1972. See 'Laying automatic submarine contact mines (Hague, VIII),' The Hague, October 18, 1907, pp.342–7.
5 John Keegan, *The Face of Battle*, New York, Viking Press, 1976, p.306.
6 With some rockets (e.g., ICBMs) we collapse this distinction and loosely call them delivery systems although these weapons are both transportation and delivery instruments strictly speaking.
7 John Keegan, 'The specter of conventional war', *Harpers*, July 1983, pp.9–14. See also James L. Foster, 'The future of conventional arms control', in John E. Endicott and Roy W. Stafford, Jr (eds), *American Defense Policy*, 4th ed., Baltimore and London, Johns Hopkins University Press, pp.127–37.
8 Laurance Martin, *Arms and Strategy: The World Power Structure Today*, New York, David McKay Company, Inc., 1973, pp.74–5.
9 Keegan, op. cit., p.10.
10 Ibid., p.10.
11 Martin, op. cit., p.79.
12 Hanson W. Baldwin, *World War I*, New York, Harper & Row, 1962, pp.156–9.
13 James F. Dunnigan, *How To Make War*, New York, Quill (William Morrow), 1983, pp.269–71.
14 Michael Walzer, *Just and Unjust Wars*, New York, Basic Books, 1977, Chapter 16.
15 *The Encyclopedia Americana*, vol.29, International Edition, Danbury, CT, Americana Corporation, 1980, p.360; Marcel Baudot, et al. (eds), *The Historical Encyclopedia of World War II*, New York, Facts On File, 1977 and 1980, pp.132–3; Andrei Sakharov, 'The danger of thermonuclear war,' *Foreign Affairs*, vol.61, no.5, 1983, p.1003. Sakharov cites estimates of deaths as the direct result of nuclear explosions into the hundreds of millions in a general war.
16 Sakharov, op. cit., p.1005–6.
17 Laurence J. Korb, *The Fall and Rise of the Pentagon: American Defense Policies in the 1970s*, Westport, CT, London, Greenwood Press, 1979, pp.144–64.
18 Dunnigan, op. cit., pp.298–9. Roughly 30% of the US warheads are land based. The comparable figure for the USSR is between 75 and 80%.
19 Ibid., pp.301–2.
20 Two volumes that discuss the unilateral armaments option extensively are *Objections to Nuclear Defence* and *Dangers of Deterrence*, both edited by Nigel Blake and Kay Pole, London, Boston, etc., Routledge & Kegan Paul, 1984.

21 Korb, op. cit., p.99.

22 Jonathan Samuel Lockwood, *The Soviet View of U.S. Strategic Doctrine*, New Brunswick, USA, London, Transaction Books, 1983, pp.7, 171–6.

23 Ronald T. Pretty (ed.), *Jane's Weapons Systems* 1981–1982, London and New York, 1982, pp.19–21.

24 Robert Jastrow, 'Reagan vs the scientists: why the president is right about missile defense,' *Commentary*, vol.78, no.1, 1984, pp.23–32.

25 Ibid., p.29.

26 John Steinbruner, 'Launch under attack,' *Scientific American*, vol.250, no.1, 1984, pp.37–47.

27 In its survey of 257 US generals and admirals *Newsweek* (July 9, 1984, p.37) reports that 21% of this group still believes that there could be a winner in a nuclear war between the US and the USSR.

28 Seymour Weiss, 'The case against arms control,' *Commentary*, vol.78, no.5, 1984, pp.19–23.

29 *Soviet Military Power*, 3rd Edition, Washington, Superintendent of Documents, US Government Printing Office, 1984, p.41. See also Weiss, op. cit., p.19.

30 Thomas H. Etzold, *Defense or Delusion? America's Military in the 1980s*, New York, Harper & Row, 1982, pp.160–2.

31 Hans A. Bethe, et al., 'Space-based ballistic-missile defense,' *Scientific American*, vol.251, no.4, 1984, pp.39–49. See especially p.47.

32 George F. Kennan, 'The way out of the nuclear dilemma,' *Bulletin of Peace Proposals*, 1981, pp.221–4. See also McGeorge Bundy, George F. Kennan, Robert S. McNamara and Gerard Smith, 'Nuclear weapons and the alliance,' *Foreign Affairs*, vol.60, no.4, 1982, pp.753–68.

33 Robert Jastrow, 'The war against "Star Wars",' *Commentary*, vol.78, no.6, 1984, pp.19–25. Jastrow disputes many of the empirical claims of the Union of Concerned Scientists (see note 31). For example, the UCS claims that 280,000 smart missiles will be needed as a part of a last-ditch defense system. Jastrow figures only 5,000 will be needed. Obviously such gross differences in numbers imply gross differences in the costs of an ABM system (p.23).

CHAPTER 9   *Civilians and the military*

1 Michael Walzer, *Just and Unjust Wars*, New York, Basic Books, 1977, Chapter 9.

2 Jeri Laber and Barnett Rubin, 'A dying nation,' *New York Review of Books*, vol.31, nos 21 & 22, 1985, pp.3–4. Laber and Rubin tell of the horrors committed by the Russians on the Afghanistan nation. The horrors include all the items listed below and more.

3 Walzer, op. cit., pp.157–9.

4 Jonathan Schell, 'The contradiction of nuclear deterrence' (from *The Fate of the Earth*), in James P. Sterba (ed.), *Morality in Practice*, Belmont, CA, Wadsworth Publishing Company, 1984, pp.324–30.

CHAPTER 10 *Guerrilla warfare*

1 W.D. Franklin, 'Clausewitz on limited war,' in Sam C. Sarkesian (ed.), *Revolutionary Guerrilla Warfare*, Chicago, Precedent Publishing, 1975, pp.179–215.
2 Stephen Goode, *Guerrilla Warfare and Terrorism*, New York and London, Franklin Watts, 1977, p.81.
3 Edward W. Gude, 'Batista and Betancourt: alternative responses to violence,' in Sarkesian, op. cit., p.582.
4 Mao Tse-Tung, *Selected Military Writings of Mao Tse-Tung*, Peking, Foreign Languages Press, 1967, p.343.
5 Robert Thompson, *Defeating Communist Insurgency*, London, Chatto & Windus, 1966, p.25.

CHAPTER 11 *Ending war*

1 Carl von Clausewitz, *On War*, ed. and intro. by Anatol Rapoport, Baltimore, MD, Penguin Books, 1972, p.101.
2 Michael Walzer, *Just and Unjust Wars*, New York, Basic Books, 1977, pp.117–20. This danger is also discussed by Stanley Hoffmann, *Duties Beyond Borders*, Syracuse, NY; Syracuse University Press, 1981, pp.50–1.
3 Walzer, op. cit., p.70.
4 Ibid., pp.117–20.
5 Michael Walzer, 'The moral standing of states: a response to four critics,' *Philosophy and Public Affairs*, vol.9, 1980, pp.213–14.
6 Walzer, 1977, op. cit., p.121.
7 Ibid., pp.113–14.
8 Charles R. Beitz, *Political Theory and International Relations*, Princeton, Princeton University Press, 1979, pp.65–6.
9 Walzer, 1977, op. cit., pp.113–14.
10 Ibid., p.114.
11 Ibid., pp.113–17.
12 Ibid., p.115.
13 Ibid., p.114.
14 Von Clausewitz, op. cit., p.119.
15 Walzer, 1977, op. cit., p.113.
16 David Luban, 'Just war and human rights,' *Philosophy and Public Affairs*, vol.9, 1980, pp.160–7.
17 Walzer, 1977, op. cit., p.112.

CHAPTER 12 *War crimes and the crime of war*

1 Richard Wasserstrom, 'On the morality of war: a preliminary inquiry,' in Malham M. Wakin (ed.), *War, Morality and the Military Profession*, Boulder, Colorado, Westview Press, 1979, p.309.
2 Michael Walzer, *Just and Unjust Wars*, New York, Basic Books, 1977, p.21.

3 Ibid., pp.114–15.
4 Richard B. Brandt, 'Utilitarianism and the rules of war,' in Malham M. Wakin, op. cit., pp.400–1.
5 Walzer, op. cit., pp.115–16.

CHAPTER 13   *Demobilization*

1 A useful examination of the costs and development of the system of veterans' benefits can be found in Sar A. Levitan and Karen A. Cleary, *Old Wars Remain Unfinished*, Baltimore and London, Johns Hopkins University Press, 1973, pp.1–14.
2 Peter Browning, 'To begin a new life,' *Nation*, vol.224, 1977, pp.39–42.
3 John M. Collins, *American and Soviet Military Trends*. Washington, D.C., The Center for Strategic and International Studies, 1978, p.169.
4 John Helmer, *Bringing the War Home*, New York, Free Press, 1974, pp.71–88, as well as Levitan and Cleary, op. cit., pp.91–4.
5 The simple analysis presented here is greatly complicated by the fact that wars responding to direct aggression are not the only morally justified wars, as we indicate in Chapters 5 and 6. Wars fought from concern for security can be analyzed in roughly the same terms, since legitimate security concerns are ultimately related to the same desire to protect life and well-being as wars responding directly to aggression. Wars fought in response to violations of human rights in other nations present greater problems. Military personnel of a nation have a duty to these aliens because of their common humanity. Once more, though, we may say that the military of the assisting nations are selected by chance, the accident that they are able to be of help and therefore have the duty to do so. There then arises the issue of what duties those assisted have to the military personnel who helped them. Often, most often, they will be unable to respond in any concrete way to the injuries suffered by their benefactors. Their response can only be gratitude. In such cases the assisting nation will have the duty to aid the suffering of its own veterans, once more, simply by the chance that it can be of help, and help is needed. Of course, assisted peoples who are able to give response to these foreign soldiers have an obligation to do so on the same grounds as if they were their own soldiers.
6 *New York Times*, Sept. 26, 1984, p.16, col.1. See also Josefina J. Card, *Lives After Vietnam*, Lexington, Mass., Lexington Books, 1983, pp.112–16.
7 J. Glenn Gray, *The Warriors*, New York, Harper & Row, 1973, pp.29–69. For a useful discussion of these problems also see Card, op. cit., pp.102–12.
8 Harold Wool, *The Military Specialist*, Baltimore, Johns Hopkins University Press, 1968, pp.93–107.

Bibliography

Abrams, Herbert and von Kaenel, William, 'Medical problems of survivors of nuclear war', *New England Journal of Medicine*, vol.305, no.20, 1981, pp.1226–32.

Appiah, Anthony, et al., 'Assessing risk', *Theory and Decision*, vol.12, 1980, pp.91–106.

Arendt, Hannah, 'Lying in politics: reflections on the pentagon papers', *New York Review of Books*, November 18, 1971, pp.30–9.

Aspaturian, Vernon V., 'The Soviet military-industrial complex: does it exist?', in Steven Rosen (ed.), *Testing the Theory of the Military Industrial Complex*, Lexington, MA, D.C. Heath, 1973.

Austin, John, *How To Do Things With Words*, New York, Oxford University Press, 1962.

Azarya. Victor, 'The Israeli armed forces', in Morris Janowitz and Stephen Westbrook (eds), *The Political Education of Soldiers*, Beverly Hills, Sage Publications Inc., 1983, pp.99–127.

Baldwin, Hanson W., *World War I*, New York, Harper & Row, 1962.

Ball, George W., 'The cosmic bluff,' *The New York Review of Books*, July 21, 1983, pp.37–41.

Baudot, Marcel, et al. (eds), *The Historical Encyclopedia of World War II*, New York, Facts On File, 1977 and 1980, pp.132–3.

Bayles, Michael D., *Professional Ethics*, Belmont, CA, Wadsworth Publishing, 1981.

Beitz, Charles R., *Political Theory and International Relations*, Princeton, Princeton University Press, 1979.

Beitz, Charles R., 'Nonintervention and communal integrity,' *Philosophy and Public Affairs*, vol.9, 1980, pp.385–91.

Bennett, John, *Nuclear Weapons and the Conflict of Conscience*, New York, Scribners, 1962.

Bennett, Jonathan, 'Whatever the consequences,' *Analysis*, vol.26, January, 1966, pp.83–102.

Bernard, Georges, 'Deterrence, utility and rational choice: a comment,' *Theory and Decision*, no.14, 1982, pp.89–97.

Blake, Nigel and Pole, Kay, *Dangers of Deterrence*, London, Boston, Routledge & Kegan Paul, 1984.

Blake, Nigel and Pole, Kay, *Objections to Nuclear Defence*, London, Boston, Routledge & Kegan Paul, 1984.

Boggs, Marion Williams, *Attempts to Define and Limit 'Aggressive' Armament in Diplomacy and Strategy*, Columbia, MO, University of Missouri Press, 1941.

Bonner, Raymond, *Weakness and Deceit: U.S. Policy and El Salvador*, New York, Times Books, 1984.

Brandt, Richard, 'Utilitarianism and the rules of war,' *Philosophy and Public Affairs*, vol.1, Winter 1972, pp.145–65.

Brandt, Richard B., *A Theory of the Good and the Right*, Oxford, Clarendon Press, 1979.

Bredow, W. von, 'The West-German Bundeswehr as an institution of political education,' in Morris Janowitz and Jacques Van Doorn (eds), *On Military Ideology*, Rotterdam University Press, 1971, pp.97–115.

Brodie, Bernard, *War and Politics*, New York, Macmillan, 1963.

Brown, Lucy, 'Intentions in the conduct of the just war,' in Cora Diamond (ed.), *Intention and Intentionality*, Ithaca, Cornell University Press, 1969.

Browning, Peter, 'To begin a new life,' *Nation*, vol.224, 1977, pp.39–42.

Bundy, McGeorge, Kennan, George F., McNamara, Robert S., and Smith, Gerard, 'Nuclear weapons and the alliance,' *Foreign Affairs*, vol.60, no.4, 1982, pp.753–68.

Calvocoressi, D., *Nuremberg: The Facts, the Law, and the Consequences*, New York, Macmillan, 1948.

Card, Josefina, J., *Lives After Vietnam*, Lexington, MA, Lexington Books, 1983.

Christoph, Bertram, *Third-World Conflict and International Security*, Hamden, CT, Archon Books, 1982.

Clausewitz, Carl von, *On War*, ed. & intro. by Anatol Rapoport, Baltimore, MD, Penguin Books, 1972.

Coady, C.A.J., 'The leaders and the led,' *Inquiry*, vol.23, 1980, pp.275–91.

Cockburn, Andrew, *The Threat: Inside the Soviet Military Machine*, New York, Vintage Books (A Division of Random House), 1983.

Cohen, M. (ed.), *War and Moral Responsibility*, Princeton, Princeton University Press, 1972.

Colby, 'War crimes,' *Michigan Law Review*, vol.23, 1925, p.606.

Collins, John M., *American and Soviet Military Trends*, Washington, DC, The Center for Strategic and International Studies, 1978.

Commission on Freedom of the Press, 'The requirements,' in Bernard Berelson and Morris Janowitz (eds), *Reader in Public Opinion and Communication*, Second Edition, New York, Free Press, 1966, pp.529–34.

Cooling, Benjamin Franklin (ed.), *War, Business and American Society: Historical Perspective on the Military-Industrial Complex*, Port Washington, NY, Kennikat Press, 1977.

Cortese, Charles F., *Modernization, Threat, and the Power of the Military*, Beverly Hills, CA, Sage Publications, 1976.

Cousins, Norman, *In Place of Folly*, revised edition, New York, Washington Square Press, 1962.

Curry, G. David, *Sunshine Patriots: Punishment and the Vietnam Offender*, Notre Dame, IN, University of Notre Dame Press, 1985.

Department of the Army Field Manual PM27–10, *The Law of Land Warfare*,

Department of the Army, July 1956.

Donagan, Alan, *The Theory of Morality*, Chicago, University of Chicago Press, 1977.

Doppelt, Gerald, 'Walzer's theory of morality in international relations,' *Philosophy and Public Affairs*, vol.8, 1978, pp.3–26.

Doppelt, Gerald, 'Statism without foundations,' *Philosophy and Public Affairs*, vol.9, 1980, pp.398–403.

Draper, Theodore, 'Nuclear temptations' and 'A postscript,' *New York Review of Books*, January 19, 1984, pp.42–50.

Draper, Theodore, 'Pie in the sky (Are the arms talks necessary?)', *New York Review of Books*, February 14, 1985, pp.20–7.

Dubik, James M., 'Human rights, command responsibility and Walzer's just war theory,' *Philosophy and Public Affairs*, vol.11, Fall 1982, pp.354–71.

Dunnigan, James F., *How To Make War: A Comprehensive Guide to Modern Warfare*, New York, Quill (William Morrow), 1982.

Dworkin, Ronald, *Taking Rights Seriously*, Cambridge, Harvard University Press, 1977.

Dworkin, Ronald, 'Is the press losing the first amendment?' *New York Review of Books*, December 4, 1980, pp.49–57.

Elfstrom, Gerard, 'On dilemmas of intervention,' *Ethics*, vol.93, 1983, pp.709–25.

Ellsberg, Daniel, 'Risk, ambiguity and rational choice,' *Quarterly Journal of Economics*, vol.75, 1961, pp.643–9.

Encyclopedia Americana, vol.29, International Edition, Danbury, CT, Americana Corporation, 1980, p.360.

Etzold, Thomas H., *Defense or Delusion? America's Military in the 1980s*, New York, Harper & Row, 1982.

Falk, Richard A., Kolko, Gabriel, and Lifton, Robert Jay (eds), *Crimes of War*, New York, Random House, 1971.

Falk, Richard A. (ed.), *The Vietnam War and International Law*, vol.III, Princeton, Princeton University Press, 1972.

Falk, Richard A., 'Environmental warfare and ecocide,' *Bulletin of Peace Proposals*, vol.4, 1973, pp.1–18.

Feemster, Cleveland W. (Sgt), 'Ethics and the noncommissioned officer,' *Ethics and the Military Profession*, 1979, pp.2–5.

Finn, James (ed.), *A Conflict of Loyalties: The Case for Conscientious Objection*, New York, Pegasus, 1968.

Ford, John C., 'The morality of obliteration bombing,' *Theological Studies*, vol.5, 1944, pp.276–85.

Ford, John C., 'The hydrogen bombing of cities,' in William J. Nagle (ed.), *Morality and Modern Warfare*, Baltimore, Helicon Press, 1960.

Foster, James L., 'The future of conventional arms control,' in John Endicott and Roy W. Stafford, Jr. (eds), *American Defense Policy*, 4th edn, Baltimore and London, Johns Hopkins University Press, 1977, pp.127–37.

Fotion, N., 'I'll bet you ten dollars that betting is not a speech act,' in Herman Parret (ed.), *Possibilities and Limitations of Pragmatics*, Philadelphia, John Benjamins North America, Inc., 1981, pp.211–23.

Franklin, W.D., 'Clausewitz on limited war,' in Sam C. Sarkesian (ed.),

Revolutionary Guerrilla Warfare, Chicago, Precedent Publishing, 1975, pp.179–215.

French, Peter A. (ed.), *Individual and Collective Responsibility*, Cambridge, MA, Schenkman Publishing Co., 1972.

Freund, Norman, 'Nonviolent national defense,' *Journal of Social Philosophy*, vol.13, May 1982, pp.12–17.

Friedman, Leon, *The Law of War: A Documentary History*, vol.1, New York, Random House, 1972.

Friedman, Milton, 'Why not a volunteer army?' in Martin Anderson (ed.), *The Military Draft*, Stanford, CA, Hoover Institute Press, 1982, pp.621–32.

Fullinwider, Robert K., 'The all-volunteer force and racial balance,' in Robert K. Fullinwider (ed.), *Conscripts and Volunteers: Military Requirements, Social Justice, and the All-Volunteer Force*, Totowa, NJ, Rowman & Allanheld, 1983, pp.178–88.

Garrison, John H., 'Military Officers and Politics II,' in John F. Reichart and Steven R. Sturm (eds), *American Defense Policy*, 5th Edition, Baltimore and London, Johns Hopkins University Press, 1982, pp.760–7.

Genovesi, Vincent J., 'Just war doctrine: a warrant for resistance,' *Thomist*, vol.45, 1981, pp.503–5.

Gewirth, Alan, *Reason and Morality*, Chicago, University of Chicago Press, 1978.

Ginzberg, Robert (ed.), *The Critique of War*, Chicago, Henry Regency, 1969.

Glossop, Ronald J., *Confronting War: An Examination of Humanity's Most Pressing Problem*, Jefferson, NC and London, McFarland, 1983.

Goode, Stephen, *Guerrilla Warfare and Terrorism*, New York and London, Franklin Watts, 1977.

Grassian, Victor, *Moral Reasoning: Ethical Theory and Some Contemporary Moral Problems*, Englewood Cliffs, Prentice-Hall, 1981.

Gray, J. Glenn, *The Warriors*, New York, Harper & Row, 1973.

Gude, Edward W., 'Batista and Betancourt: alternative responses to violence,' in Sam C. Sarkesian (ed.), *Revolutionary Guerrilla Warfare*, Chicago, Precedent Publishing, 1975, pp.569–85.

Gutteridge, William, *Military Institutions and Power in the New States*, New York, Frederick A. Praeger, Inc., 1965.

Halle, Louis, J., 'Does war have a future?', *Foreign Affairs*, vol.52, no.1, 1973, pp.20–34.

Hare, J.H. and Joynt, Carey B., *Ethics and International Affairs*, New York, St Martin's Press, 1982.

Hare, R.M., *Freedom and Reason*, Oxford, Oxford University Press, 1965.

Hare, R.M., 'Rules of war and moral reasoning,' *Philosophy and Public Affairs*, vol.1, 1972, pp.166–81.

Hare, R.M., *Moral Thinking*, Oxford, Clarendon Press, 1981.

Harries-Jenkins, Gwyn, 'The British armed forces,' in Morris Janowitz and Stephen Westbrook (eds), *The Political Education of Soldiers*, Beverly Hills, Sage Publications Inc., 1983, pp.83–93.

Hastings, Max and Jenkins, Simon, *The Battle of the Falklands*, London and Sydney, Pan Books, 1983.

Heilbroner, Robert, 'Military America,' *New York Review of Books*, July 23, 1970, pp.5–8.

Held, Virginia, Morgenbesser, Sidney and Nagel, Thomas (eds), *Philosophy, Morality, and International Affairs*, New York, Oxford University Press, 1974.

Heller, Mark (ed.), *The Middle East Military Balance*, 1983, Boulder, CO, Westview Press, 1983.

Helmer, John, *Bringing the War Home*, New York, Free Press, 1974.

Herz, John H., 'The survival problem,' in Melvin Kransberg (ed.) *Ethics in an Age of Pervasive Technology*, Boulder, CO, Westview Press, 1980.

Higgs, Calvin T. Jr. (Captain), 'Military professionalism and the emergence of the NCO,' *Ethics and the Military Profession*, 1979, pp.5–18.

Hoffmann, Stanley, *Duties Beyond Borders*, Syracuse, NY, Syracuse University Press, 1981.

Hoffmann, Stanley, 'States and the morality of war,' *Political Theory*, vol.9, 1981, pp.149–72.

Hunter, Richard W. and Nelson, Gary R., 'The all-volunteer force: has it worked, will it work? (Summary)', in Martin Anderson (ed.), *Registration and the Draft*, Stanford, CA, Hoover Institute Press, 1982, pp.11–20.

Huntington, Samuel P., *The Soldier and the State*, Cambridge, MA, Harvard University Press, 1957.

Huntington, Samuel P., *The Common Defense*, New York, Columbia University Press, 1961.

Janowitz, Morris, *The Professional Soldier*, Glencoe, IL, Free Press of Glencoe, 1960.

Janowitz, Morris, 'The logic of national service,' in Martin Anderson (ed.), *The Military Draft*, Stanford, CA, Hoover Institute Press, 1982, pp.403–27.

Jaspers, Karl, *The Question of German War Guilt*, New York, Dial, 1947.

Jastrow, Robert, 'Reagan vs the scientists: why the president is right about missile defense,' *Commentary*, vol.78, no.1, 1984, pp.23–32.

Jastrow, Robert, 'The war against "Star Wars",' *Commentary*, vol.78, no.6, 1984, pp.19–25.

Jervis, Robert, *Perception and Misperception in International Politics*, Princeton, Princeton University Press, 1976.

Johnson, James T., *Just War Tradition and the Restraint of War: A Moral and Historical Inquiry*, Princeton, Princeton University Press, 1982.

Johnson, James T., 'The moral bases of contingency planning,' *Hastings Center Report*, vol.12, April, 1982, pp.19–20.

Johnson, Chaplain Kermit D. (Colonel), 'Ethical issues of military leadership,' *Parameters*, vol.4, 1974, pp.35–9.

Kahn, Herman, *On Thermonuclear War*, Princeton, Princeton University Press, 1960.

Kaplan, Mark, 'Rational acceptance,' *Philosophical Studies*, vol.40, Sept. 1981, pp.129–46.

Karas, Thomas, *The New High Ground: Systems and Weapons of Space Age War*, New York, Simon & Schuster, 1983.

Kavka, Gregory, 'Deterrence, utility, and rational choice,' *Theory and Deci-*

sion, vol.12, March 1980, pp.41–60.

Kavka, Gregory, 'Deterrence and utility again: a response to Bernard,' *Theory and Decision*, vol.14, March 1982, pp.99–102.

Keegan, John, *The Face of Battle*, New York, Viking Press, 1976.

Keegan, John, 'The specter of conventional war,' *Harpers*, July 1983, pp.9–14.

Kelsen, H., 'Collective and individual responsibility in international law with particular regard to the punishment of war criminals,' *California Law Review*, vol.31, 1953.

Kennan, George F., 'The way out of the nuclear dilemma,' *Bulletin of Peace Proposals*, 1981, pp.221–4.

Kennedy, Edward M., 'Inequities in the draft,' in Martin Anderson (ed.), *The Military Draft*, Stanford, CA, Hoover Institute Press, 1982, pp.527–9.

Knoll, Erwin, and Nies, McFadden, Judith Nies (eds), *War Crimes and the American Conscience*, New York, Holt, Rinehart & Winston, 1970.

Koistinen, Paul A.C., *The Military Industrial Complex: A Historical Perspective*, New York, Praeger, 1980.

Korb, Laurence J. *The Fall and Rise of the Pentagon: American Defense Policies in the 1970s*, Westport, CT, London, Greenwood Press, 1979.

Laber, Jeri and Rubin, Barnett, 'A dying nation,' *New York Review of Books*, January 17, 1985, pp.3–4.

Lackey, Douglas P., 'Ethics and nuclear deterrence,' in James Rachels (ed.), *Moral Problems*, New York, Harper & Row, 1979.

Lackey, Douglas P., 'Missiles and morals: a utilitarian look at nuclear deterrence,' *Philosophy and Public Affairs*, vol.11, 1982, pp.189–231.

Laird, Melvin R., 'People, not hardware: the highest defence priority,' *American Enterprise Institute Special Analysis*, Washington, DC, American Enterprise Institute.

Lapp, Ralph E., *The Weapons Culture*, New York, W.W. Norton, 1968.

Lefever, Ernest W., 'Facts, calculation, and political ethics,' in *The Moral Dilemma of Nuclear Weapons: Essays from Worldview*, New York, Council on Religion and International Affairs, 1961.

Lens, Sidney, *The Military Industrial Complex*, Philadelphia, Pilgrim Press, 1970.

Levinson, Sanford, 'Responsibility for crimes of war,' *Philosophy and Public Affairs*, vol.2, 1973, pp.244–73.

Levitan, Sar A. and Cleary, Karen A., *Old Wars Remain Unfinished*, Baltimore and London, Johns Hopkins University Press, 1973.

Lewey, G., 'Superior orders, nuclear warfare, and the dictates of conscience,' *American Political Science Review*, vol.55, March 1961, pp.3–23.

Lockwood, Jonathan Samuel, *The Soviet View of U.S. Strategic Doctrine*, New Brunswick, USA, London, Transaction Books, 1983.

Luban, David, 'Just war and human rights,' *Philosophy and Public Affairs*, vol.9, 1980, pp.161–81.

Luban, David, 'The romance of the nation-state,' *Philosophy and Public Affairs*, vol.9, 1980, pp.392–7.

MacIntyre, Alasdair, *After Virtue*, Notre Dame, University of Notre Dame Press, 1981.

Malamet, David, 'Selective conscientious objection and the Gilette decision,' *Philosophy and Public Affairs*, vol.1, 1972, pp.363–86.

Mao Tse-Tung, *Selected Military Writings of Mao Tse-Tung*, Peking, Foreign Languages Press, 1967.

Marlowe, David H., 'The meaning of the force and the structure of the battle: Part I – The AVF and the draft,' in Robert K. Fullinwider (ed.), *Conscripts and Volunteers: Military Requirements, Social Justice, and the All-Volunteer Force*, Totowa, NJ, Rowman & Allanheld, 1983, pp.46–57.

Martin, Laurance, *Arms and Strategy: The World Power Structure Today*, New York, David McKay Company, Inc., 1973.

Mathews, Robert, 'National security, propaganda or legitimate concern,' in Asbjorn Eide and Marek Thee (eds), *Problems of Contemporary Militarism*, New York, St Martin's Press, 1980.

Mavrodes, George I., 'Conventions and the Morality of War,' *Philosophy and Public Affairs*, vol.4, Winter 1975, pp.117–31.

Mead, Margaret, 'A national service system as a solution to a variety of national problems,' in Martin Anderson (ed.), *The Military Draft*, Stanford, CA, Institute Press, 1982, pp.431–43.

Melman, Seymour, *Pentagon Capitalism*, New York, McGraw Hill, 1970.

Menzel, Paul T. (ed.), *Moral Argument and the War in Vietnam: A Collection of Essays*, Nashville, TN, Aurora Publishers, 1971.

Merleau-Ponty, Maurice, *Humanism and Terror*, Boston, Beacon Press, 1969.

The Military Balance 1980–1981, London, The International Institute for Strategic Studies, 1980.

Milovidov, A.S. and Zhdanov, E.A., 'Sociophilosophical problems of war and peace,' *Soviet Studies in Philosophy*, vol.20, 1981, pp.3–39.

Monro, D.H., 'Civil rights and conscription,' in Martin Anderson (ed.), *The Military Draft*, Stanford, CA, Hoover Institute Press, 1982, pp.133–51.

Moskos, Charles C., Jr., *The American Enlisted Man*, New York, Russell Sage Foundation, 1970.

Moskos, Charles C., Jr. (ed.) *Public Opinion and the Military Establishment*, Beverly Hills, CA, Sage Publications, 1970.

Moskos, Charles C., Jr., 'The military-industrial complex: theoretical antecedents and conceptual contradictions,' in Sam Sarkesian (ed.), *The Military-Industrial Complex: A Reassessment*, Beverly Hills, CA, London, Sage Publications Inc., 1972, pp.3–23.

Moskos, Charles C., Jr., 'The all-volunteer force,' in Morris Janowitz and Stephen D. Westbrook (eds), *The Political Education of Soldiers*, Beverly Hills, CA, London and New Delhi, Sage Publications, 1983, pp.307–25.

MQS I Training Support Package: Ethics and Professionalism, Fort Benjamin Harrison, IN, The Military Professional Ethics Division, Training Development Directorate, US Army Support Center, 1981.

MQS II Training Support Package: Ethics and Professionalism, Fort Benjamin Harrison, IN, The Military Professional Ethics Division, Training Development Directorate, US Army Support Center, 1982.

MQS III Training Support Package: Ethics and Professionalism, Fort Benjamin

Harrison, IN, The Military Professional Ethics Division, Training Development Directorate, US Army Support Center, 1983.

Murray, John Courtney, *Morality and Modern War*, New York, Church Peace Union, 1960.

Nagle, William J. (ed.), *Morality and Modern Warfare*, Baltimore, Helicon Press, 1960.

Narel, James (Captain), 'What is the ethical code of the noncommissioned officer?', *Ethics and the Military Profession*, 1979, pp.19–22.

Narveson, Jan, 'Pacifism: a philosophical analysis,' *Ethics*, vol. 75, 1965, pp.259–71.

Neilands, J.B., et al., *Harvest of Death: Chemical Warfare in Vietnam and Cambodia*, New York, Free Press, 1972.

Nincic, Miroslav, *The Arms Race: The Political Economy of Military Growth*, New York, Praeger Publishers, 1982.

Nolan, Richard T. and Kirkpatrick, Frank G. (with Harold H. Titus and Morris T. Keeton), *Living Issues in Ethics*, Belmont, CA, Wadsworth, 1981.

O'Brien, William V., *War and/or Survival*, Garden City, NY, Doubleday, 1969.

Oi, Walter, Y., 'The economic costs of the draft,' in Martin Anderson (ed.). *The Military Draft*, Stanford, CA, Hoover Institute Press, 1982, pp.317–46.

Orlow, Alexander, *Handbook of Intelligence and Guerrilla Warfare*, Ann Arbor, MI, University of Michigan Press, 1963.

Paskins, Barrie and Dockrill, Michael, *The Ethics of War*, Minneapolis, MN, University of Minnesota Press, 1979.

Paulson, Stanley, L., 'Classical legal positivism at Nuremberg,' *Philosophy and Public Affairs*, vol.4, Winter 1975, pp.132–58.

Pretty, Ronald T. (ed.), *Jane's Weapons Systems*, 1981–1982, London and New York, James, 1982.

Prisoner-of-War-Resistance, Field Manual No 21–78, Washington, DC, Headquarters, Department of the Army, 1981.

Ra'anan, Uri, Pfaltzgraff, Robert L. Jr., and Kemp, Geoffrey (eds), *Projection of Power*, Hamden, CT, Shoe String Press, 1982.

Rains, Roger A. (Captain), and McRea, Michael J. (Captain) (eds), *The Proceedings of the War and Morality Symposium*, West Point, NY, United States Military Academy, 1980.

Ramsey, Paul, *War and the Christian Conscience*, Durham, NC, Duke University Press, 1961.

Ramsey, Paul, *The Just War: Force and Political Responsibility*, New York, Scribners, 1968.

Ramsey, Paul, 'How shall counter-insurgency war be conducted justly?', in Paul T. Menzel, *Moral Argument and the War in Vietnam: A Collection of Essays*, Nashville, TN, Aurora Publishers, 1971.

Raskin, Marcus G. and Fall, Bernard B. (eds), *The Vietnam Reader*, New York, Vintage Books, 1965.

Rawls, John, *A Theory of Justice*, Cambridge, MA, Harvard University Press, 1971.

Regan, Tom, 'A defense of pacifism,' *Canadian Journal of Philosophy*, vol.2, 1972, pp.73–86.

Richards, David, A.J., 'Rights, resistance and the demands of self-respect,' *Emory Law Journal*, vol.32, 1983, pp.405–35.

Royal Commission on the Press, 'The standard by which the press should be judged,' in Bernard Berelson and Morris Janowitz (eds), *Reader in Public Opinion and Communication*, Second Edition, New York, Free Press, 1966, pp.535–42.

Ryan, Cheney C., 'Self-defense, pacifism and the possibility of killing,' *Ethics*, vol.93, 1983, pp.508–24.

Russell, Bertrand, 'Bertrand Russell on negotiations,' *New Republic*, vol.138, 1958, p.9.

Sakharov, Andrei, 'The danger of thermonuclear war,' *Foreign Affairs*, vol.61, no.5, 1983, pp.1001–16.

Scanlon, Thomas, 'Contracturalism and utilitarianism' in Amartya Sen and Bernard Williams (eds), *Utilitarianism and Beyond*, Cambridge, Cambridge University Press, 1982, pp.103–28.

Schauer, Frederick, *Free Speech: A Philosophical Enquiry*, Cambridge, Cambridge University Press, 1982.

Schell, Jonathan, 'The contradiction of nuclear deterrence,' (from *The Fate of the Earth*), in James P. Sterba (ed.), *Morality in Practice*, Belmont, CA, Wadsworth Publishing Company, 1984, pp.324–30.

Schelling, Thomas C., *Arms and Influence*, New Haven and London, Yale University Press, 1966.

Schmidt, Steffen W., *El Salvador, America's Next Vietnam?*, Salisbury, NY, Documentary Publications, 1983.

Schwarzenberger, Georg, *The Legality of Nuclear Weapons*, London, Stevens, 1958.

Scolnick, Joseph M., Jr., 'Case studies – Britain and Canada,' in James C. Miller, III (ed.), *Why the Draft?* Baltimore, MD, Penguin Books Inc., 1968, pp. 91–104.

Searle, John, 'A taxonomy of illocutionary acts,' *Expression and Meaning*, Cambridge, Cambridge University Press, 1979, pp.1–29.

Sharp, Gene, *Social Power and Political Freedom*, Boston, Sargent, 1980.

Shawcross, William, *Sideshow*, New York, Simon & Schuster, 1979.

Shue, Henry, *Basic Rights*, Princeton, Princeton University Press, 1980.

Simpson, H.B., 'Compulsory military service in England,' in Martin Anderson (ed.), *The Military Draft*, Stanford, CA, Hoover Institute Press, 1982, pp.463–77.

Slater, Jerome, 'Military officers and politics I,' in John F. Reichart and Steven R. Sturm (eds), *American Defense Policy*, 5th Edition, Baltimore and London, Johns Hopkins University Press, 1982, pp.749–56.

Smith, T.V., 'Ethics for soldiers of freedom,' *Ethics*, vol.60, 1950, pp.157–68.

Somerville, John, 'Democracy and the problem of war,' in Paul Kurtz (ed.), *Moral Problems in Contemporary Society*, Englewood Cliffs, NJ, Prentice-Hall, 1969.

Somerville, John, 'Patriotism and war,' *Ethics*, vol.9, 1981, pp.568–78.

Soviet Military Power, 3rd Edition, Washington, Superintendent of Documents, US Government Printing Office, 1984.

Stawinski, Jansz, 'On the ubiquity of violence,' *Dialectics and Humanism*, vol.7, 1980, pp.165-75.

Stein, Walter (ed.), *Nuclear Weapons: A Catholic Response*, London, Merlin Press, 1965.

Steinbruner, John, 'Launch under attack,' *Scientific American*, vol.250, no.1, 1984, pp.37-47.

Swomley, J.M., *The Military Establishment*, Boston, Beacon Press, 1964.

Tahtnen, Dale R., *National Security Challenges to Saudi Arabia*, Washington, DC, American Enterprise Institute for Public Policy Research, 1975.

Taylor, A.J.P., *A History of the First World War*, New York, Berkeley, 1963.

Taylor, Telford, 'Just and unjust wars,' in Malham M. Wakin (ed.), *War, Morality and the Military Profession*, Boulder, CO, Westview Press, 1979, pp.245-58.

Thee, Marek, 'Militarism and militarization in contemporary international relations,' in Asbjorn Eide and Maret Thee (eds), *Problems of Contemporary Militarism*, New York, St Martin's Press, 1980, pp.15-35.

Thompson, Kenneth W., *The President and the Public Philosophy*, Baton Rouge, Louisiana State University Press, 1981.

Thompson, Robert, *Defeating Communist Insurgency*, London, Chatto & Windus, 1966.

Thomson, Judith Jarvis, 'Self-defense and human rights,' *The Lindley Lectures*, Lawrence, KA, University of Kansas Press, 1976.

Tucker, Robert W., *The Just War*, Baltimore, Johns Hopkins University Press, 1960.

Union of Concerned Scientists, 'Reagan's star wars,' *New York Review of Books*, April 26, 1984, pp.47-52.

United States Congress, Office of Technology Assessment, *The Effects of Nuclear War*, Montclair, NJ, Allanheld, Osmun, 1979.

Upton, Emory (Brevet Major General), *The Military Policy of the United States*, New York, Greenwood Press, 1968, first published 1904.

Wakin, Malham M., *War, Morality and the Military Profession*, Boulder, CO, Westview Press, 1979.

Waldman, Eric, *The Goose Step is Verboten: The German Army Today*, New York, Free Press, 1964.

Walzer, Michael, 'Moral judgment in time of war,' *Dissent*, vol.14, 1967, pp.284-92.

Walzer, Michael, *Obligations: Essays on Disobedience, War, and Citizenship*, New York, Simon & Schuster, 1971.

Walzer, Michael, 'World War II: why was this war different?', *Philosophy and Public Affairs*, vol.1, 1971, pp.3-21.

Walzer, Michael, *Just and Unjust Wars*, New York, Basic Books, 1977.

Walzer, Michael, 'The moral standing of states: a response to four critics,' *Philosophy and Public Affairs*, vol.9, 1980, pp.209-29.

Wasserstrom, Richard A., 'Three arguments concerning the morality of war,' *Journal of Philosophy*, vol.65, 1968, pp.578-90.

Wasserstrom, Richard A., 'On the morality of war: a preliminary inquiry,'

Stanford Law Review, vol.21, 1969, pp.1627–56.

Wasserstrom, Richard A. (ed.), *War and Morality*, Belmont, CA, Wadsworth, 1970.

Wasserstrom, Richard A., 'The relevance of Nuremberg,' *Philosophy and Public Affairs*, vol.1, 1971, pp.22–46.

Wasserstrom, Richard A., 'The laws of war,' *The Monist*, vol.56, 1972, pp.1–19.

Wasserstrom, Richard A., 'Book Review: Walzer: *Just and Unjust Wars*,' *Harvard Law Review*, vol.92, 1978, pp.536–45.

Wasserstrom, Richard A., 'On the morality of war: a preliminary inquiry,' in Malham M. Wakin (ed.), *War, Morality and the Military Profession*, Boulder, CO, Westview Press, 1979, pp.299–325.

Wasserstrom, Richard A., *Philosophy and Social Issues*, Notre Dame, IN, University of Notre Dame Press, 1980.

Weisberg, Barry, *Ecocide in Indochina*, San Francisco, Canfield Press, 1970.

Weiss, Seymour, 'The case against arms control,' *Commentary*, vol.78, no.5, 1984, pp.19–23.

Wolff, Robert Paul, 'On violence,' *Journal of Philosophy*, vol.66, 1969, pp.601–16.

Wool, Harold, *The Military Specialist*, Baltimore, Johns Hopkins University Press, 1968.

World Armaments and Disarmament: Stockholm International Peace Research Institute Yearbook 1981, London, Taylor & Francis, 1981.

World Armaments and Disarmament: Stockholm International Peace Research Institute Yearbook 1983, New York, Taylor & Francis, 1983.

Zinn, Howard, *Disobedience and Democracy: Nine Fallacies on Law and Order*, New York, Random House, 1968.

Zoll, Ralph, 'The German armed forces,' in Morris Janowitz and Stephen D. Westbrook (eds), *The Political Education of Soldiers*, Beverly Hills, CA, Sage, 1983, pp.209–48.

Index

government officials, 234; for
war crimes, 245, 258, 279
Tupemaro guerrillas, 214–16

UCMJ, 80
United Nations, 116, 254
United States, 24, 29, 31, 40, 41, 42,
50, 52, 54, 55, 56, 60, 89, 101,
102, 117, 119, 120, 121, 122,
123, 126, 175–6, 178–9, 182,
183, 184, 190–1, 227, 248, 249,
251, 253, 260, 262; Veterans'
Administration, 262
universal military service, 55–7
universalizability, 13, 17, 21, 150;
principle of, 150, 207
unstable military conditions, 176–8,
181–91, 209–11
Uruguay, 214–17
USSR, 29, 31, 40, 42, 102, 117,
175–6, 178–9
utilitarianism, 1, 8, 10–11, 12–14,
15–21, 22–4, 36, 46, 93–4, 136,
206, 210–11, 280
utilitarians, 53, 200

Venezuela, 216
veterans, 262–76
Viet Cong, 227
Vietnam, 73, 112, 119, 123, 227,
253
virtue-based morality, 281–2
virtues, 78
volunteer military force, 49–57, 60

Walzer, Michael, 9–10, 110–12,
124–5, 155, 172, 228, 229–43,
255–6
war, 6–10, 12, 22–3, 24, 25, 37, 54,
109–10, 123, 192–3, 220, 221,
230–1, 277; civil, 278; the crime
of, 244–61; crimes, 232, 235,
244–61; defining features of, 2;
goals of, 227–32; guerrilla,
212–24, 278; just, 23, 107–8,
129, 160, 203, 237, 250, 253,
272–3, 277; justification of, 23,
39, 107–18; military goals of,
236–43; political goals of,
236–43; pre-emptive strike in,
117; wrongful, 245–9, 275
warfare, area, 205, 206, 212, 223,
258
Washington, George, 262
weapons: area, 159, 167, 169,
196–205, 214; biological, 169,
173–4, 258; nuclear, 54, 159,
174–91, 208–10
Wilson, Woodrow, 125, 239
World War I, 125, 168, 169, 174,
199, 227, 239
World War II, 22, 41, 45, 54, 56,
111, 128, 155, 156, 163, 168,
174, 199, 206, 219, 220, 221,
227, 231, 235, 236, 255, 274

Yugoslavia, 219

For Product Safety Concerns and Information please contact our EU
representative GPSR@taylorandfrancis.com
Taylor & Francis Verlag GmbH, Kaufingerstraße 24, 80331 München, Germany

www.ingramcontent.com/pod-product-compliance
Lightning Source LLC
Chambersburg PA
CBHW070557270326
41926CB00013B/2345